29 Advances in Polymer Science

Fortschritte der Hochpolymeren-Forschung

Edited by H.-J. CANTOW, Freiburg i. Br. · G. DALL'ASTA, Colleferro
K. DUŠEK, Prague · J. D. FERRY, Madison · H. FUJITA, Osaka
M. GORDON, Colchester · W. KERN, Mainz · G. NATTA, Milano
S. OKAMURA, Kyoto · C. G. OVERBERGER, Ann Arbor · T. SAEGUSA, Kyoto
G. V. SCHULZ, Mainz · W. P. SLICHTER, Murray Hill · J. K. STILLE, Fort Collins

With 76 Figures

Springer-Verlag
Berlin Heidelberg GmbH 1978

Editors

ISBN 978-3-662-15454-0 ISBN 978-3-540-35795-7 (eBook)
DOI 10.1007/978-3-540-35795-7

Library of Congress Catalog Card Number 61-642

Originally published by Springer-Verlag Berlin Heidelberg New York in 1978.
Softcover reprint of the hardcover 1st edition 1978

2152/3140 – 543210

Contents

Polymers with Photoconductive Properties

Milan Stolka and Damodar M. Pai

Xerox Corporation, Rochester, N.Y. 14644 U.S.A.

Photoconductivity of organic polymers is reviewed. The current views on the mechanism of charge carrier generation and transport in sensitized and unsensitized polymers are presented. The known photoconductive polymers are categorized according to their structure into three basic groups: polymers with high degree of conjugation in the main chain, polymers with condensed aromatic groups and polymers with substituted arylamino functional groups.

Table of Contents

A. Introduction

Photoconductivity in a solid is defined as an increase of conductivity caused by radiation. The phenomenon of photoconductivity involves the processes of absorption of radiation, photogeneration of charge carriers, their separation, diffusion and drift in an applied electric field, their temporary immobilization at sites known as trapping sites, release from traps and finally their recombination. The phenomenological relationships covering all these processes were primarily developed in connection with the study of crystalline covalent solids which dominated the early scientific literature on photoconductivity. Concurrent with the basic understanding of the phenomena was the devolopment of several experimental techniques to study the fundamental processes and the specific identity of the defects and impurities that control these processes.

Although significant advances have been made in the understanding of photoconductivity in organic molecular crystals, the studies of organic polymeric systems have lagged behind. This was partly caused by the complexities of the disordered states. The organic polymeric systems differ from the crystalline states in that they are further characterized by several material parameters such as molecular weight and its distribution, presence of end groups, purity etc. Besides, until very recently the technological applications, which are generally a strong motivating force behind a substantial research effort were lacking. Photoconductivity phenomena in the organic and polymeric systems have several unusual features, primarily the result of weak interactions between the molecules as opposed to the covalent crystalline solids where transport occurs through allowed energy bands. The absorption of light in organic and polymeric systems is represented by the discrete transitions between the ground state and the excited electronic states of the active species in the macromolecule. Therefore intramolecular relaxation and electron-phonon interactions play a dominant role. The photogeneration and charge transport properties in such situations are best described by analogy with chemical oxidation and reduction reactions in which the intermolecular charge transfer involves molecular ions and neutral molecules.

The highly insulating nature of most of the organic polymers coupled with the low photogeneration efficiency and charge carrier mobilities resulting from the disorder present formidable experimental problems in that the standard semiconductor techniques such as steady state dc photoconductivity, Hall effect and thermopower measurements are not easily applicable. Nevertheless photoconductivity in many polymers discussed in the literature has been characterized by employing the steady state photoconductivity technique without any consideration regarding the nature of the electrode contacts and the build up of space charge within the sample. This aspect coupled with the lack of essential data on the light source employed makes it difficult to accurately assess how photoconductive some of these polymers really are. In this context we would like to paraphrase a statement made by Rose[1)] in the introduction of his book "Concepts in Photoconductivity and Allied Problems" – "The dimensions of the field of photoconductivity are so impressive . . one would have to make an effort to find materials in which photocurrents are not detectable."

However, there are certain experimental techniques that are uniquely suited to the study of highly insulating photoconductors with low charge carrier mobilities. In the first part of this chapter we review some basic concepts of photoconductivity which are followed by a review of some experimental techniques and how these have been applied to characterize some of the well known polymeric systems such as poly(N-vinyl carbazole) (PVK) and the charge transfer complex of PVK and 2,4,7,trinitro-9-fluorenone (TNF). The second part of this chapter is a review of the extensive original and patent literature on a variety of photoconducting polymers.

B. Some Basic Concepts in Photoconductivity

I. Definitions

Let us consider a solid subjected to bulk absorbing radiation producing g_0 carriers/cm^3 sec. The steady state concentration of carriers is given by

$$n_0 = g_0 \tau$$

where τ, the recombination time, is the time interval between the photogeneration of the carriers and the time at which they disappear by recombination. Throughout this review for the sake of simplicity we will consider the case where only one carrier is mobile. The steady state conductivity upon irradiation is given by

$$\sigma = n_0\, e\mu = g_0 \tau e\mu$$

where e is the electronic charge and μ is the drift mobility. n_0 is the value of the carrier density before the solid is connected in any circuit. However, most of the traditional measurement techniques require some sort of contacts and the very act of connecting the solid across a voltage source may result in a change of the carrier density and the measured current has no relation to the fundamental parameters n_0, μ and τ. Only when the contacts are ohmic, the steady state carrier density remains unchanged after the sample is connected in a measuring circuit. This contact assures that when a carrier leaves the solid at one contact another carrier enters the solid at the opposite contact to maintain the carrier density at n_0. It has to be emphasized that ohmic contacts are not easily made on these systems and are an exception rather than a rule. The photocurrent with ohmic contacts is

$$\begin{aligned} i_s &= \sigma E = e g_0 \tau E \mu \\ &= e\left(\frac{\eta I}{L}\right) \tau E \mu \end{aligned} \qquad (1)$$

where I is the total absorbed radiation in photons/cm^2 sec, η, the quantum efficiency (QE) of generation, is the number of carriers produced per absorbed photon and L is the film thickness. Since the proportionality coefficient connecting i_s with E contains η, μ and τ, Eq. (1) shows why the phenomenon of photoconductivity has been used to measure the basic transport parameters in solids. It is normally necessary to conduct other experiments to disentangle the various parameters in Eq. (1). The current under ohmic contact conditions varies linearly with the electric field only when η, μ and τ are constants. However, in all the polymeric systems characterized to date these quantities vary with the electric field and therefore the ohmic current variation with the field is a very complicated function.

On the other hand the measured current upon irradiation by the same source would be different if the solid has blocking contacts. Perfect blocking contacts are defined as those that permit the photogenerated carriers to drift out of the solid under the influence of the electric field but do not allow any charge to be injected into the solid. The current under these conditions is given by[2].

$$i = \eta I e \tag{2}$$

The tacit assumption here is that the recombination time τ between oppositely charged carriers is larger than the transit time T_t of the carriers so that the measured current is equal to the rate at which carriers are being produced (multiplied by e to convert the photon current into electrical current). Any electric field dependence of this current then reflects the electric field dependence of the QE of photogeneration. However, in the presence of substantial carrier recombination ($\tau < T_t$) the current with blocking contacts is

$$i_p = e\eta I \frac{\tau}{T_t} = e I \eta \frac{\mu E}{L} \tag{3}$$

which is identical to the expression for the current under ohmic conditions. In the presence of substantial recombination the probability that a photogenerated carrier arrives at an electrode is small compared to the probability of its recombination with the opposite carrier. As a consequence, only few carriers arrive at the electrode and the contacts do not have to replenish carriers. The current i_p under blocking contact conditions is sometimes referred to as the primary current and the ratio

$$\frac{i_s}{i_p} = \frac{\tau}{T_t} = G$$

is sometimes referred to as the gain[1]. It equals the number of transits which a photogenerated carrier makes along the length of the specimen before it recombines with the opposite carrier. The steady state photosensitivity is therefore determined by the quantum efficiency of photogeneration η, the recombination time τ and the transit time T_t. The steady state photosensitivity can be increased by increasing η and τ and reducing T_t (or increasing μ).

The mobility μ is the velocity of the charge carriers per unit electric field and can be defined as either the microscopic mobility μ_0 or the drift mobility μ_d. The microscopic mobility is the mobility of the carriers in the absence of any traps. For example, in organic molecular systems it may be the mobility as determined by hopping between molecular states. However, if one introduces trapping sites into the solid, the carrier in addition to hopping between the states is also temporarily immobilized in trapping sites and the net velocity of the carriers is reduced. This reduced velocity per unit field is defined as the drift mobility. Obviously in the absence of trapping states the microscopic and the drift mobilities have identical values.

II. Experimental Techniques

It is clear that the steady state photoconductivity measurement [Eq. (1)] does not provide any information on the fundamental parameters η, μ and τ of the photoconductor even under the assumption that ohmic contacts can be made. These parameters can be best studied by transient techniques. In principle these techniques involve creation of a sheet of carriers whose transport across the polymer film is time resolved. From the shape and the magnitude of the current pulse, it is possible to obtain information on the photogeneration efficiency, transit times (drift mobility) and trapping characteristics[2]. The polymer film sandwiched between a conducting substrate and a semitransparent metallic top electrode is connected in series with a voltage source and a resistance R (Fig. 1). If the polymer is insulating in the dark the entire voltage V_0 appears across the film. The device is now irradiated through the semitransparent electrode by a light pulse of total flux F. If the absorption coefficient is high, the light is absorbed in a thin polymer region close to the top electrode. The RC time constant circuit is made large compared to the transit time of the carriers T_t. In this open circuit case the potential across the film drops when the photogenerated carriers (ηF) drift to the substrate. An equal and opposite voltage signal appears across the resistance. Its value is

$$\begin{aligned} \Delta V(t) &= \frac{\Delta Q(t)}{C} = \frac{\eta Fe}{C}\frac{t}{T_t} && \text{for } t < T_t \\ &= \frac{\eta Fe}{C} && \text{for } t > T_t \end{aligned} \tag{4}$$

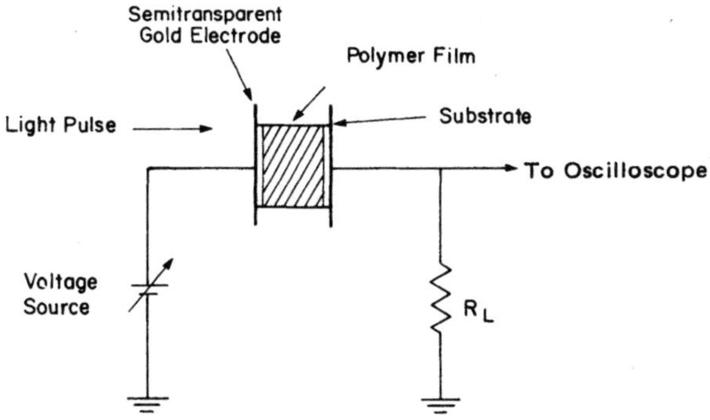

Fig. 1. Schematic of the experimental arrangement used for measuring transient photoconductivity

In the derivation of Eq. (4) it is assumed that the charge in motion ηFe is small compared to the charge on the electrode CV_0. It is also assumed that the charge pulse moves across as a sheet with uniform velocity $\frac{L}{T_t}$ without any carrier getting immobilized by traps.

If on the other hand the carriers are lost by bulk trapping, $\Delta V(t)$ becomes,

$$\Delta V(t) = \frac{\eta F}{C} \frac{\tau_t}{T_t} [1-\exp(-t/\tau_t)] \text{ for } t < T_t \qquad (5)$$

where τ_t is the free time after which the carrier is immobilized in traps. It is called trapping time. The amplitude of the pulse at $t = T_t$ under heavy bulk trapping $(\tau_t < T_t)$ is given by

$$\Delta V = \frac{\eta F}{C} \tau_t \frac{\mu E}{L} \qquad (6)$$

The voltage pulse which is a ramp of amplitude $\frac{\eta Fe}{C}$ in the absence of trapping, is an exponential with an amplitude $\frac{\eta F}{C} \tau_t \frac{\mu E}{L}$ under heavy bulk trapping conditions. The above experiment is ideally suited to determine the extent of trapping in a polymer film. In the absence of trapping at constant applied field the amplitude of the pulse (in coulombs) is independent of the thickness of the film. On the other hand it is inversely proportional to the sample thickness in the presence of trapping. If it is determined that the bulk trapping is not a factor, the amplitude of the pulse is proportional to the QE of photogeneration.

If the experiment is carried out with a small series resistance R such that $RC < T_t$ (short circuit case) the current in the absence of trapping is

$$\begin{aligned} i = \frac{e\eta F}{T_t} &= e\eta F \frac{\mu E}{L} \quad \text{for } t < T_t \\ &= 0 \qquad\qquad \text{for } t > T_t \end{aligned} \qquad (7)$$

The signal in this case is a rectangle of amplitude $e\,\eta F \frac{\mu E}{L}$. This experiment is popularly known as the time of flight technique and has been successfully employed in measurements of charge carrier mobility and QE of photogeneration in photoconductive polymers. If a particular polymer has a very low QE of photogeneration it is possible to inject carriers into the film from an external source and measure the charge carrier mobility. This external source could be a thin overcoating of a sensitive photoconductor such as amorphous selenium or the carriers could be excited in the metal electrode and injected into the polymer film (internal photoemission) by choosing a light source of appropriate wavelength.

A variation of the experimental methods described above is the xerographic discharge technique which is gaining wide acceptance in the study of polymeric systems. The polymer film deposited on a metallic substrate is corona charged and irradiated by a light source of appropriate wavelength and known intensity. The polarity of the corona charge is chosen to be the same as the sign of the majority carrier. The potential difference across the film is monitored by a capacitively coupled probe. In the absence of trapping the rate of change of potential is related to the QE of photogeneration by the relation[2],

$$C = \frac{dV}{dt} = -\eta I e \tag{8}$$

This expression is identical to that of the steady state current obtained for an electroded sample with blocking electrodes. The tacit assumption made in the derivation of Eq. (8) is that the total charge drifting through the sample is less than the initial charge on the surface of the film CV_0 and the rate of change of potential is measured at times larger than the transit time. It is also assumed that the surface is blocking to corona species and the substrate contact is non injecting. This experiment can be used to measure photogeneration efficiencies in polymeric films.

Though the time of flight technique described earlier is widely used in measurements of charge carrier mobilities in low mobility polymers, the transient current signals observed in many systems are not easily interpreted. In such cases a variation of the xerographic discharge technique performed under space charge limited conditions (SCLC) provides an indirect method of obtaining information on the charge carrier mobility and its electric field dependence. A corona charged film of a highly absorbing polymer deposited on a conducting substrate is exposed to radiation. The intensity of the source is arranged to drive the film under SCLC. The SCLC condition prevails if the maximum permissible charge CV_0 is in transit. The initial discharge rate (at $t \simeq T_t$) under these conditions is given by [2,3],

$$\frac{dV}{dt} = \frac{-V_0}{2} \frac{1}{T_t} \tag{9}$$

Once again, the experiments are best done on overcoated polymer films where the overcoating consists of a thin layer of sensitive photoconductor so that CV_0 of charge can be photogenerated with reasonable light intensity values.

C. PVK and PVK : TNF

I. Photogeneration

A knowledge of the photo-physical processes taking place in the organic molecular state is an essential prerequisite to the understanding of the photogeneration process. The excited molecular state can dissipate its energy by fluorescence, phosphorescence, and radiationless deactivation and therefore these processes are competitive with photocarrier generation. The circumstances and factors involved in photogeneration and transport will be illustrated in the case of poly(N-vinylcarbazole) (PVK). In Fig. 2

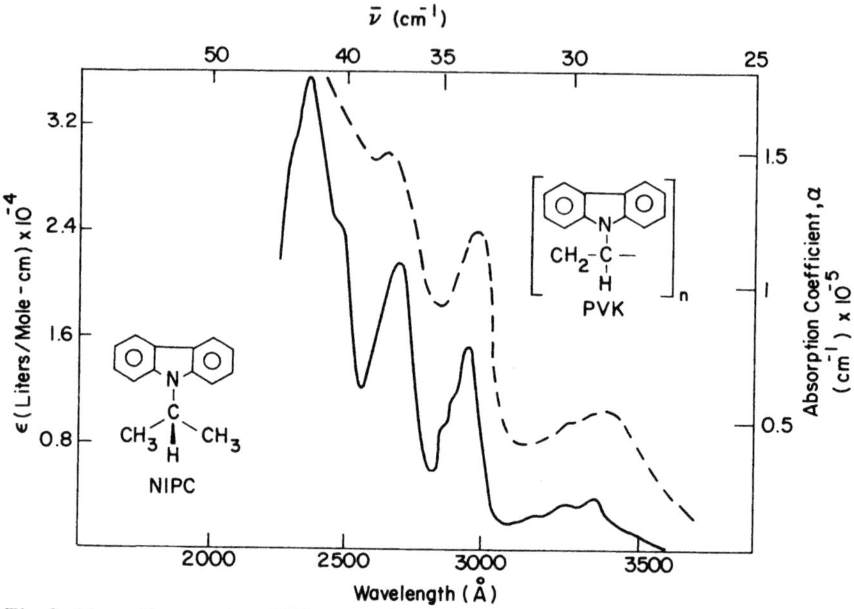

Fig. 2. Absorption spectra of N-isopropylcarbazole in dichloromethane (free molecule) (solid line) and a thin film of the polymer, polyvinylcarbazole (dotted line)[4)]

the absorption spectrum of PVK in solid state is seen to be remarkably similar to that of the free molecule. The structural similarity of the two arises from the weak van der Waals interaction between individual molecules in a solid state.

According to Regensburger[5)], the photogeneration spectrum had a similar structure. A model in which singlet excitions interact with impurity centers after migrating to the surface would explain why the photogeneration spectrum is similar in shape to the absorption spectrum. However in some recent studies, the efficiency of carrier generation was found to increase in a stepwise fashion[6)] with decreasing wavelength. The steps occur at 3600 A°, 3100 A°, 2850 A° corresponding to the electronic states of the carbazole molecule. The absolute QE at 2540 A° measured in PVK films of varying thicknesses are plotted in Fig. 3 as a function of the applied electric field[7)]. These measurements were carried out on electroded samples using the time of flight technique in which the circuit time constant is made larger than the transit time of the carriers. The collected charge at a constant field is seen to be independent of film thickness. This indicates that what is measured is in fact the charge injected into the sample from the thin (~0.1 μ) photoabsorption region near the top surface. The electric field dependence of the collected charge therefore represents the field dependence of the photogeneration processes which occur in the absorption region *i.e.* at the surface. The most successful model employed to date to explain the field dependence of photogeneration is that proposed by Onsager[8)].

According to this model an absorbed photon produces a hole and electron pair which as a result of the finite distance between them, experiences a substantial coulombic attraction. As a result of this attractive force a fraction of these pairs

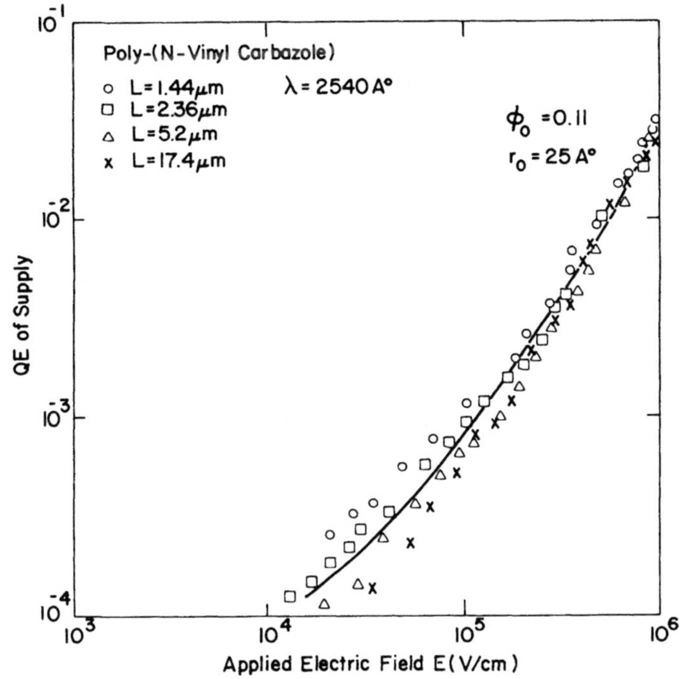

Fig. 3. Circles, squares, triangles and crosses are the QE of supply as a function of electric field for PVK films of different thickness. The solid line is the Onsager expression for $\phi_0 = 0.11$ and $r_0 = 25$ A°

recombine and the remainder dissociate. In the presence of an applied field, this dissociation process is enhanced at the expense of the recombination process and the efficiency of the dissociation is increased. The recombination between the hole and electron created by the same photon is called geminate or initial recombination.

The theory of geminate recombination reduces to the problem of Brownian motion in the presence of Coulomb attraction. Onsager's relationship for the probability $p(r, \theta, E)$ that an ion pair (thermalized with an initial separation r and at an angle θ with the applied field direction) will escape recombination is given by[9)],

$$p(r,\theta,E) = e^{-A}e^{-B} \sum_{n=0}^{\infty} \sum_{m=0}^{\infty} \frac{A^m}{m!} \frac{B^{m+n}}{(m+n)!} \tag{10}$$

where $A = \dfrac{e^2}{4\,\pi\epsilon\epsilon_0 kTr}$ and $B = \dfrac{eEr}{2\,kT}(1 + \cos\theta)$

and e is the electronic charge, $\epsilon\epsilon_0$ is the absolute dielectric constant of the medium, k the Boltzman constant and E the electric field. By defining ϕ_0 as the efficiency of production of thermalized ion pairs per absorbed photon, and $g(r,\theta)$ as the initial separation between ions of each ion pair, we arrive at the formula for the overall generation efficiency

$$\phi(E) = \phi_0 \int p(r,\theta,E)\, g(r,\theta)\, d^3r \tag{11}$$

where ϕ_0 is assumed to be independent of the applied field. We also assume that the initial distribution of thermalized pairs is an isotropic δ function, so that

$$g(r,\theta) = \frac{1}{4\pi r_0^2}\,\delta(r - r_0) \tag{12}$$

where r_0 is a characteristic thermalization length. After integrating Eq. (11) the resulting expression for the escape or the photogeneration efficiency is given[10,11] by

$$\phi(r_0, E) = \phi_0 \frac{kT}{eEr_0} e^{-A} e^{-eEr_0/kT} \sum_{m=0}^{\infty} \frac{A_0^m}{m!} \sum_{n=0}^{\infty} \sum_{\ell=m+n+1}^{\infty} \left(\frac{eEr_0}{kT}\right)^{\ell} \frac{1}{\ell!} \tag{13}$$

where $A_0 = \dfrac{e^2}{4\pi\epsilon\epsilon_0 kTr_0}$

On expansion, the first few terms of the expression can be written as

$$\phi(r_0, E) = \phi_0 \left\{\exp \frac{-r_c(T)}{r_0}\right\} \left\{1 + \frac{e}{kT}\frac{1}{2!} r_c E + \left(\frac{e}{kT}\right)^2 \frac{1}{3!} r_c \left(\frac{r_c}{2} - r_0\right) E^2 + \ldots\right. \tag{14}$$

where r_c is the critical Onsager distance at which the Coulomb energy is equal to kT and is defined by

$$r_c(T) = \frac{e^2}{4\pi\epsilon\epsilon_0 kT}$$

The solid line in Fig. 3 shows the values calculated from the Onsager expression with the maximum photogeneration efficiency ϕ_0 (at infinite field) of 0.11 and r_0 of 25 A°. The good agreement clearly establishes that in the final step the carrier pairs produced by only 11 percent of the absorbed light undergo Onsager type dissociation process. What remains to be established now is which one of the many excited states – excimer, exciplex, charge transfer state or the normal singlet/triplet manifold – of the molecules participates in the dissociation process.

An excimer is an excited dimer state obtained by the interaction between the excited singlet state of the molecule with a neutral molecule. Excimer formation involves the delocalization of the electron between the two molecules. Purified PVK films show broad featureless excimer fluorescence with a maximum around 4000 A°[12]. If the excimers were to undergo Onsager dissociation, the excimer flourescence would decrease with increasing field. This is not observed[13]. On the other hand, if the excited state in the singlet/triplet manifold were to undergo Onsager dissociation, the resulting efficiency would be extremely small because of the very small separation distance between the oppositely charged carriers. In addition to that it would be hard to explain the r_0 of 25 A° obtained by fitting the Onsager expression to the measured photogeneration efficiency. The dissociation of charge transfer states on the other hand is a possibility. The formation of the charge transfer state may or may not involve an acceptor type impurity[14],

$$D^* + D \rightarrow D^+ + D^- \text{ (charge transfer state)}$$

$$D^* + A \rightarrow (D^+ \ldots A^-)^* \rightarrow D^+ + A^-$$

$$A^- + D \text{ or } D^* \rightarrow D^- + A$$

where D is the ground state of PVK, D* is the singlet excited state, A is an electron acceptor type impurity such as residual solvent or another impurity or possibly dissolved oxygen or oxidized carbazole units, $(D^+ \ldots A^-)^*$ is an exciplex. The formation of a charge transfer state by autoionization followed by its Onsager type dissociation may also be the mechanism of photogeneration in purified PVK. In commercial PVK or PVK containing acceptor type impurities, exciplex formation followed by its dissociation results in free carrier production.

A similar mechanism has been proposed[15] to explain photogeneration data in the 1 : 1 charge transfer complex of PVK and 2,4,7-trinitro-9-fluorenone (TNF). The data points in Fig. 4 show the measured photogeneration efficiency and the solid line is calculated from the Onsager expression with $\phi_0 = 0.23$ and $r_0 = 35$ A°. For a

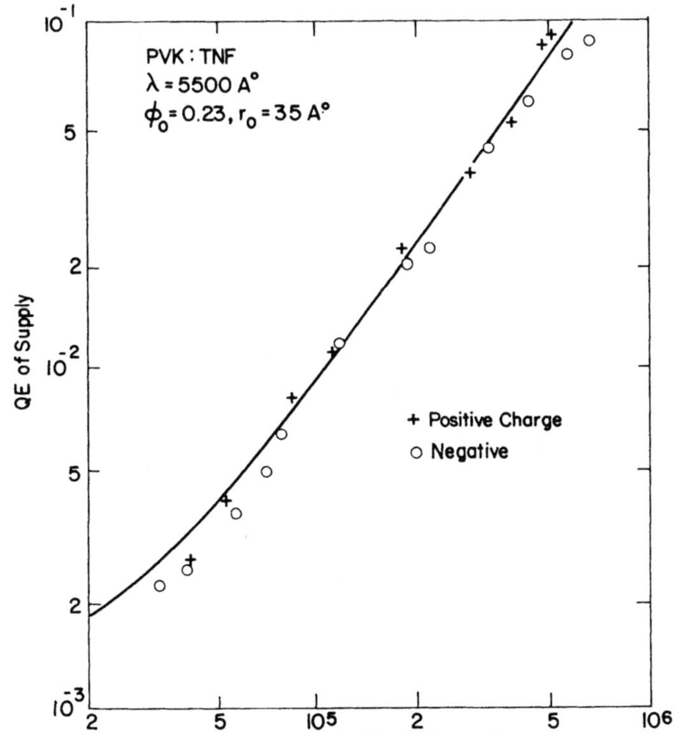

Fig. 4. Photogeneration efficiency of 1 : 1 molar ratio TNF to PVK monomer as a function of applied field at 23 °C. Theoretical computation uses $r_0 = 35$ A° and $\phi_0 = 0.23$[15]

0.06 : 1 TNF :PVK combination the corresponding values were 0.23 and 25 A° respectively. The invariance of ϕ_0 with increased TNF concentration is explained by assuming that ϕ_0 is determined by local processes within a TNF-PVK (monomer) complex. The increase in r_0 is attributed to the extended nature of the electronic states which influence the autoionization process.

II. Charge Transport

Charge transport through organic polymeric systems shows some unusual features. When the time of flight experiments are performed in inorganic crystalline solids the charge carriers drift in a sheet without any dispersion (except for the normal diffusion effects). All the carriers exit the sample at a specific time T_t. However a similar experiment with polymer films shows a very dispersive transit (Fig. 5 a) which indicates that only a small fraction of the carriers exit the sample at $t = T_t$. A large number of carriers remain within the sample for more than two or three times the transit time of the fastest carriers. A smaller fraction remain within the sample for considerably longer times. In addition the charge carrier mobility in amorphous polymer systems increases with the applied electric field whereas in crystalline materials it remains invariant.

The shape of the transient current pulses in disordered systems has been the subject of intense debate in the past few years. In the absence of distinct conduction bands, transport proceeds by hopping through a network of localized sites. The fiduciary point or the cusp in the transient current pulse corresponds to a small fraction of the carriers that cross the sample within a certain time without encountering any "long hops". At that time, however, a large fraction of the carriers encounter long hops whose hopping time might be of the order of magnitude of, or even larger than, the transit time of the fastest carriers. These carriers will eventually cross the sample giving rise to the long tail. The transient current was found to decay algebraically with two power law variations of $t^{-(1-\alpha)}$ followed by $t^{-(1+\alpha)}$ with the value of α between 0 and 1 [16]. The point of intersection between these two lines on a log i – log t plot is the transit time of the fastest carriers. This method of determination of transit times is employed in the case of extremely dispersive transport when the transient current pulse does not have a clearly defined cusp as shown in Fig. 5 (a).

The electric field dependence of the hole mobility in a series of PVK : TNF films of varying composition is shown in Fig. 5 (b)[17]. The carrier drift mobilities are extremely low and strongly dependent on the electric field. Similar electric field dependence was observed for electron drift mobilities. The mobilities obey the empirical relation

$$\mu = \mu_0 \exp\left[-(E_0 - \beta E^{1/2})/kT_{eff}\right] \tag{15}$$

where $\dfrac{1}{T_{eff}} = \dfrac{1}{T} - \dfrac{1}{T_0}$

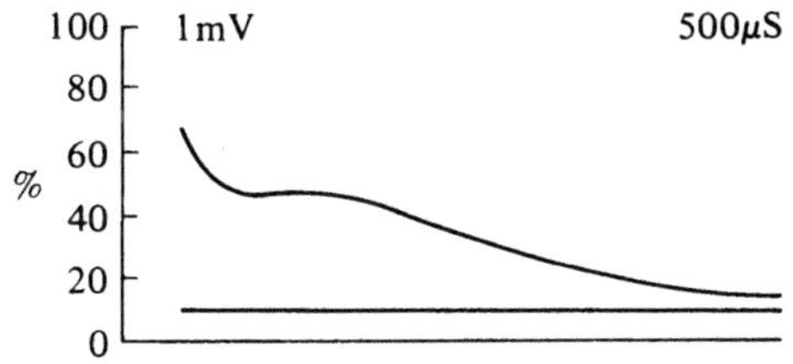

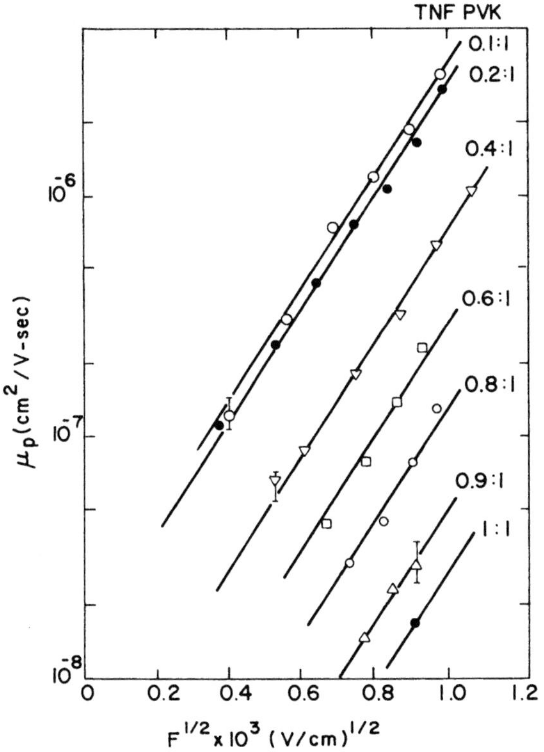

Fig. 5. Field dependence of hole drift mobilities for a range of TNF : PVK molar ratios. Data taken at T = 24 °C[17)]

where the prefactor mobility μ_0, the activation energy E_0 and a temperature T_0, obtained by fitting the field and temperature dependence to the measured data, vary with the film composition (See Table 1). β is found to be $2.7 \times 10^{-5}\ \mathrm{eV}\left(\frac{\mathrm{V}}{\mathrm{m}}\right)^{-1/2}$ and is the same for films of all compositions.

From the absorption and electro-absorption experiments it was shown[18)] that the degree of complexing varies with the composition of the film. The absorption in the visible region saturates above a ratio of 0.6 : 1 TNF : PVK. The complexing varies from about 72% at 0.1 : 1 TNF : PVK to 40% at 1 : 1 which indicates that TNF : PVK is a three-component system containing free TNF, uncomplexed carbazole and CT complexes. The transport in PVK is dominated by holes and in pure

Table 1. Parameters in Eq. (15) obtained for various TNF : PVK molar ratios[17)]

TNF : PVK	Holes			Electrons		
	E_0(eV)	T_0(K)	μ_0(cm^2/V.s)	E_0(eV)	T_0(K)	μ_0(cm^2/V.s)
0 : 1	0.65	660	2×10^{-2}	–	–	–
0.1 : 1	0.65	625	1×10^{-2}	0.67	595	1.6×10^{-5}
0.2 : 1	0.68	521	2.7×10^{-3}	0.71	568	3×10^{-4}
0.4 : 1	0.68	519	4.6×10^{-4}	0.68	545	1×10^{-4}
0.6 : 1	0.65	550	3.8×10^{-4}	–	–	–

TNF it is dominated by electrons. Figure 6 shows the variation of the hole and electron drift mobility with the TNF : PVK molar ratio. This plot provides provides valuable information on the nature of the transport in these materials[17)]. The figure reveals that the hole transport is associated with uncomplexed carbazole, electron

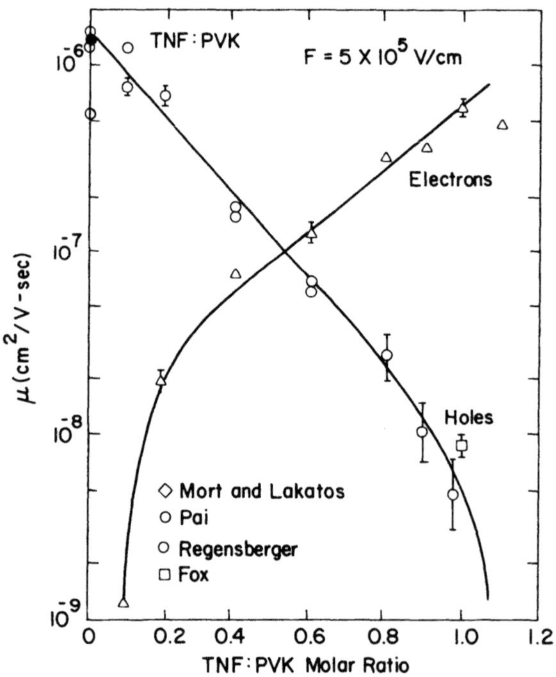

Fig. 6. Variation of hole and electron drift mobility with TNF : PVK molar ratio[17)]

transport with both free and complexed TNF. The extremely low value of the mobility coupled with the steep electric field dependence indicates a high degree of localization. It also shows that the charge transport proceeds *via* an intermolecular hopping process. The reduction of the hole mobility with increasing TNF concentration results from the increase in the average distance between the monomer units of uncomplexed PVK.

The empirical relationship [Eq. (15)] obtained by fitting the experimental electric field dependence to the measured data suggests a Poole-Frenkel (PF) type mechanism[17]. The PF effect deals with a carrier trapped in a coulombic barrier[19]. The applied field causes a reduction of $\beta_{PF}E^{1/2}$ in the barrier height and therefore the probability of escape is increased in the presence of the applied field. The slope $\beta_{PF}E^{1/2}$ predicted by the theory is $\left(\frac{e}{\pi\epsilon\epsilon_0}\right)^{1/2}$. The calculated value is 4.08×10^{-5} eV$(V/m)^{-1/2}$ for an ϵ of 3.5 as compared to the experimental value of 2.72×10^{-5}. The PF theory can be improved by incorporating both the details of the shape of the coulomb potential and the microscopic diffusion mechanism of the charge escape[19]. It is then possible to derive an expression whose slope is lower than β_{PF} and which is therefore in a better agreement with the measured β.

Several models have been proposed[17] to explain the origin and details of the PF mechanism operative in these materials. One is the traditional trap controlled drift mobility mechanism where the trap depth is modulated by the applied field. This would require a large number of charged coulombic centers compensated by an equal number of oppositely charged centers to maintain the neutrality of charge. Since the slope of the field dependence remains invariant with the differing ratios of PVK and TNF it is difficult to rationalize the presence of these charged centers in materials ranging from PVK, through the charge transfer complex of PVK:TNF at all molar ratios to pure TNF.

Another mechanism that has been proposed is that the carriers move as small polarons[20]. A small polaron is a carrier that is "self trapped" in a well created by the lattice distortion. This lattice distortion is formed when a carrier stays sufficiently long in a position to polarize the medium around it. The applied field can lower the polaron barrier in a PF fashion and increase the mobility. The polaron transport model is attractive in that the mobilities in this mechanism are not critically dependent on the sample preparation.

D. Photoconductive Polymers

As discussed above, the phenomenon of photoconduction involves two distinctly different steps: photogeneration of charge carriers and their transport. Both steps can be intrinsic qualities of the polymer, especially in the UV range. If a polymer has a capability of both producing and transporting charge carriers it is properly called a photoconductor. The carriers can however be generated in another adjacent material and injected into the polymer. In such a case the polymer should not be called a photoconductor but merely an electronic conductor or a charge carrier transport polymer. The term photoconductor has been often used too liberally in descriptions of polymers which clearly provided only the transport function. The classification of polymers to photoconductive and transporting should be related to the particular wavelength of interest. Unsensitized PVK, for example is a true photoconductor in the UV range but only a transporting medium in the visible range where an extrinsic carrier generator is needed to accomplish photoconduction.

From the structural point of view the known photoconductive (or transporting) polymers can be divided into three groups:

1) Polymers with a high degree of conjugation in the main chain. In these polymers the absorption extends well into the visible and in some cases into the near infrared region, especially when certain electron withdrawing groups are present. They are photoconductors in the visible region of the spectrum.
2) Polymers with pendant or inchain large polynuclear aromatic groups with large π electron systems. These polymers usually absorb only UV light below 400 nm in which range they are photoconductive. Outside the UV range they act as carrier transport materials.
3) A large and diversified group of polymers containing a substituted aromatic amine group in many configurations, either pendant or in-chain. Again, unless the structure contains a strong electron withdrawing group, the polymers are photoactive only below ~400 nm but are useful as carrier transport media in conjunction with a carrier generator active above 400 nm. Most polymers which are reported to be useful in electrophotography belong to this category.

I. Polymers with Polyconjugated Systems of Multiple Bonds

Polymers and oligomers with conjugated $>C=C<$, $>C=N-$ and –N=N-double and –C≡C– triple bonds in the main chain received recently some attention for their photoconductive properties. Although most of these polymers are infusible and intractable, some of them are reported to possess good solubility and film-forming properties and outstanding thermal stability.

Among the first polyconjugated polymers found to display an internal photoeffect were polyquinazones such as polyacenequinone radicals and aniline black[21)]

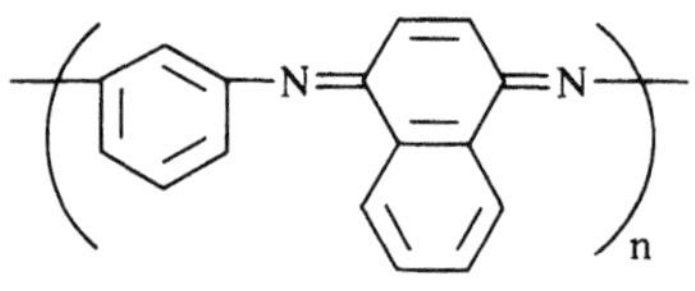

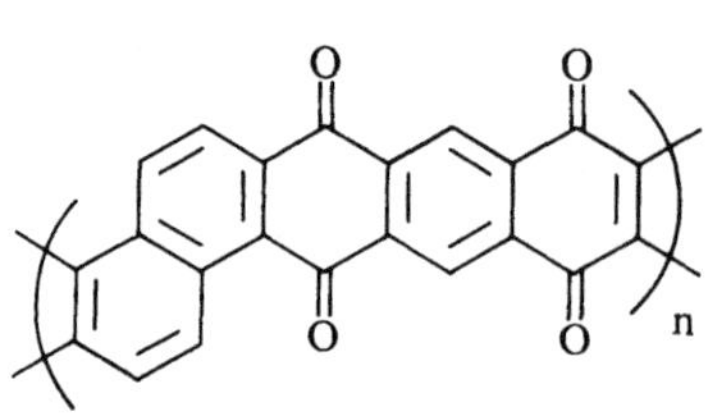

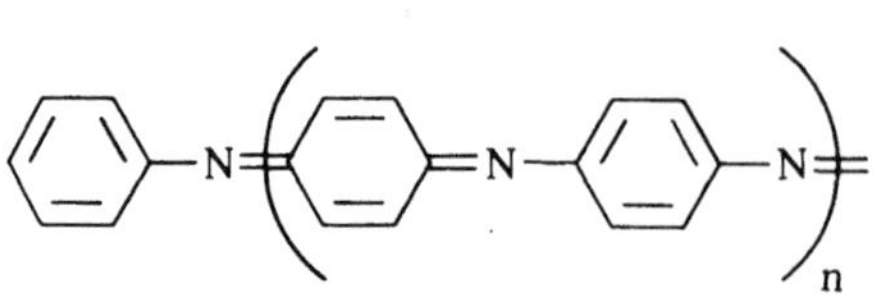

and polyynes, *i.e.* polymers with conjugated C≡C triple bonds[22]. Mylnikov[23] reviewed the data obtained with this type of polymer until 1973 and found that photocurrents i are proportional to the light intensity I according to the formula $i = I^n$ where $0.5 < n < 1$; n becomes unity at long wavelengths[23,24]. Holes are the majority carriers in all cases. Mylnikov and others concluded that photogeneration of charge carriers involves formation of excitons followed by their dissociation in an electric field during an interaction with a defect or another exciton. The activation energies of photoconduction are between 1.5 and 2.5 eV which is similar to the activation energies of dark conduction, suggesting that both photo and dark processes are of the same nature.

Chance and Baughman[24] studied photoconductivity in poly[2,4-hexadiyn-1,6-diol bis(p-toluene sulfonate)][25]. Simple polyacetylenes such as this one are unique

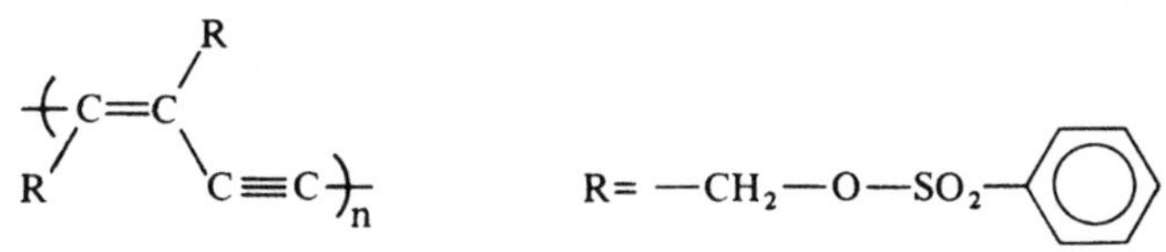

among polyconjugated polymers in that they can be obtained as single crystals and therefore they have been considered as suitable for fundamental studies.

The ratio of peak photocurrent to dark current was as high as 300 at 300 °K and 600 at 120 °K. The action spectrum of photoconductivity of this and other studied polydiacetylenes[24] is significantly shifted to shorter wavelengths from the absorption spectrum.

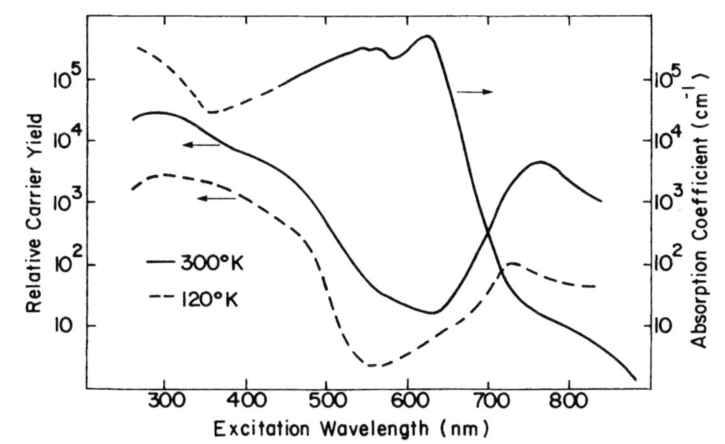

Fig. 7. Relative carrier yield versus excitation wavelength. All data are normalized to the same absorbed photon flux. The electric field was ~1300 V/cm oriented approximately in the polymer chain direction in a standard surface cell configuration. The peak photocurrent-to-dark current ratio was ~300 at 300 °K and >600 at 120 °K. Also shown is the absorption spectrum of PTS. The solid portion of the curve refers to the polymer chain direction. The dashed portion is an extrapolation derived from absorption measurements for a dispersion of PTS in a KBr pellet[24]

The authors concluded that the excited states created by the optical transition at peak absorption (~620 nm) are not involved in carrier generation. Very interesting is the observation that photocurrents increased with decreasing temperature to the peak value at 175–225 °K. They proposed that carrier mobilities increase with decreasing temperature and that the increase is high enough to offset the exponential decrease in carrier separation and lifetime.

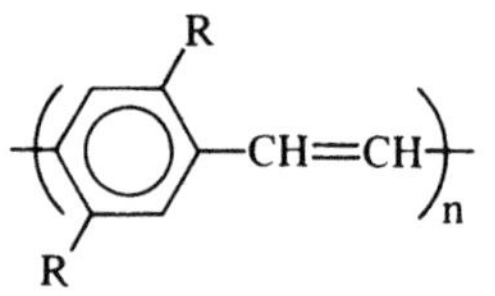

R = H; CH_3; $O{-}CH_3$

Horhold[26] observed that the activation energy E_{phot} of photoconduction in poly-p-xylylidenes is also near 2.35 eV, which is the same as E_{dark} (2.4 eV). The absorption and photoconductivity action spectra are, however, equivalent. The photoconductivity in an oxygen atmosphere is larger than in argon; the difference varies with the nature of the substituent R. Oxygen apparently plays a role in the carrier generation.

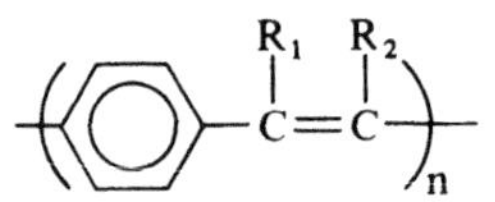

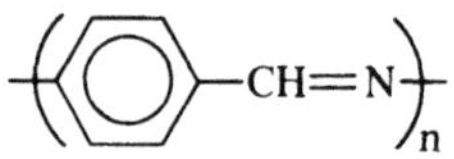

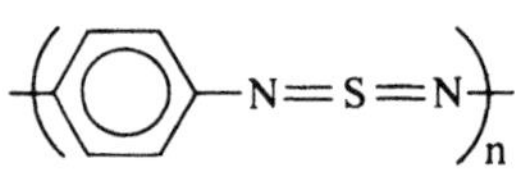

A number of other substituted polyarylenes such as polyxylylidenes poly(aryleneazomethines) and polyarylenesulfurdiimides were found to be photoconductive with similar properties as the above polymers[27–30]. The polyarylenesulfurdiimide based on benzidine exhibits a spectral response into the infrared region[27]. The substituents X and Y as well as the structure of the aromatic group have a profound influence on the photoconductivity and other physical properties of the polymers. The interruption of the polyconjugation when the aryl group is

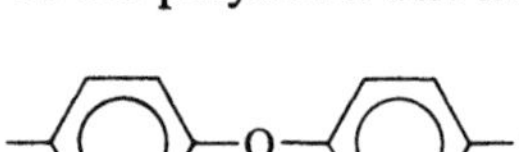 or *meta* 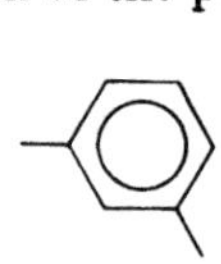 results in a loss of photoconductivity[28].

Hörhold found that the action spectrum of phenyl-substituted polyarylenevinylenes shifts substantially to longer wavelengths when the photoconductor probe is simultaneously irradiated by UV light from another source[31]. The action spectrum of photoconductivity of the polymer subjected to UV radiation is the same as the action spectrum of photoconductivity of the cation radical of the same polymer generated chemically by oxidation with antimony pentachloride. He assumed that the photoconduction involves photochemical generation of cation radicals (by UV) and optical excitation (by visible or near infrared light) of the macromolecular cation radical.

This effect is different from that observed earlier by Mylnikov[32] who found that preirradiation by UV enhances the photoconductivity of metal polyacetylenides. Mylnikov proposed that the bond rupture and trapping of electrons is responsible for the enhancement.

Another group of polyconjugated photoconductive polymers reported in the literature are poly(Schiff-bases)[33], polycondensates of aromatic diamines and

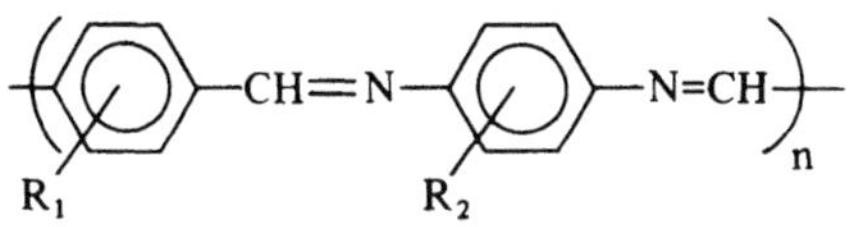

aromatic dialdehydes with very low degree of polymerization (n = 2–5). The model compounds representing the monomer units did not exhibit photoconductivity. In similar experiments, the highest photosensitivity was observed with poly(Schiff-bases), terminated with arylamino groups[34]. In general, however, the poly(Schiff-bases) are quite inefficient photoconductors; the observed quantum efficiency of generation is only 5×10^{-6} to 1×10^{-4} electrons per photon[34].

Another group of polymers and oligomers with extended conjugation in the chain, polystyrylpyrimidines[35] with the structure suggested as follows were identified as fairly efficient photoconductors. The ratio of photocurrent to dark current

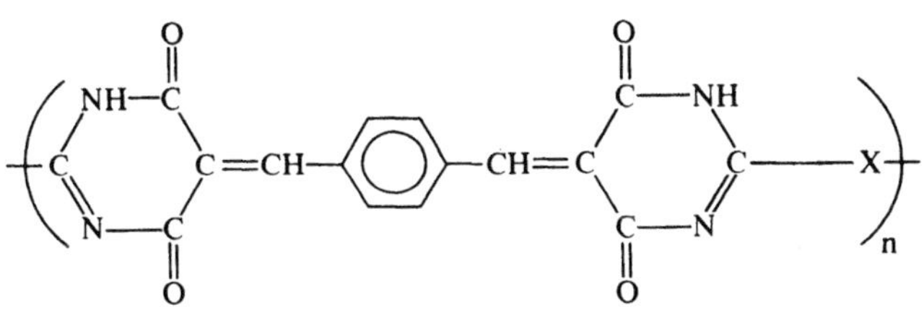

X =—OCO— Aryl—COO—

at peak photoconductivity was up to 7300. The photocurrent values reached up to $\sim 7 \times 10^{-11}$ A at electric fields of 10^2 V/cm. The photoconductivity also depends on atmosphere; the system becomes less efficient in oxygen than it is in an inert atmosphere. This and similar observations show how unreliable the data on photogeneration can be if such an important factor as the presence of oxygen is not considered in the analysis of the results.

Some photoconductivity was observed in polymers of dimethyl and diethyl-acetylene dicarboxylate[36].

Reucroft *et al.*[37,38] reported photoconductivity of polymers with polyconjugation in the chain, prepared by polycondensation of 3,3′-diaminobenzidine and 3,3,′4,4′-benzophenone tetracarboxylic acid dianhydride[39], pyrrone polymers:

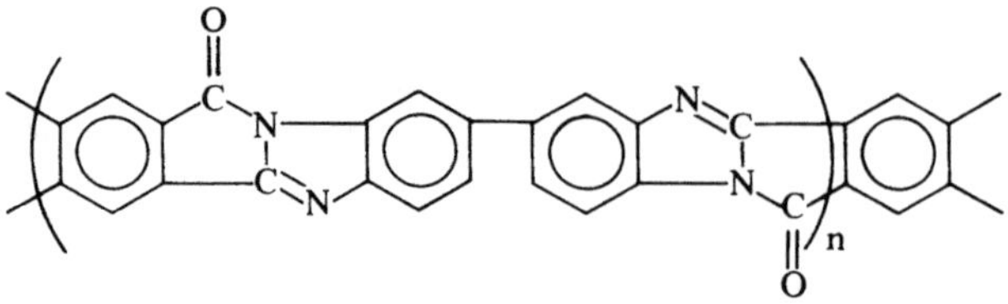

The ratio of photo- to dark current was as high as 200–300 with panchromatic light intensity of 20–30 W/cm^2; the photocurrents reached 10^{-7} A/cm^2 at electric fields of 4–5 x 10^5 V/cm. In these polymers the absorption spectra coincide with the photoconductivity action spectra.

Similar photoconductive polymers were later reported by Voishchev *et al.*[40] with the structure proposed as follows:

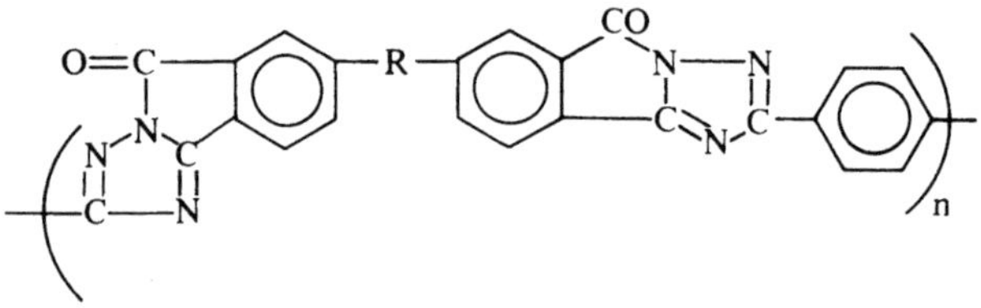

These s.c. poly[(benzoylene)-s-triazoles] have poorly defined structure since the cyclization reaction is incomplete. Since the photoconductivity is higher when a substantial amount of unreacted amino groups are left in the structure these polymers and oligomers could be classified as polymers with photoconductive amino group. Voishchev *et al.*[41] recently reported photoconductive properties of poly-[N-phenyl)benzimidazoles] such as

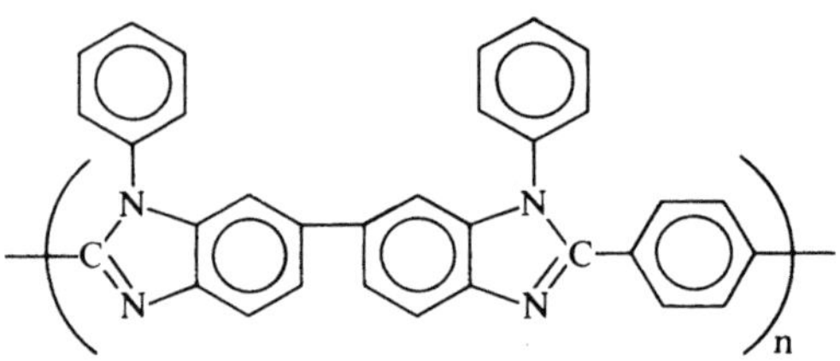

with peak response at 390–410 nm. Holes were identified as majority carriers.

Similarly, poly(benzoxazoles)[42] such as

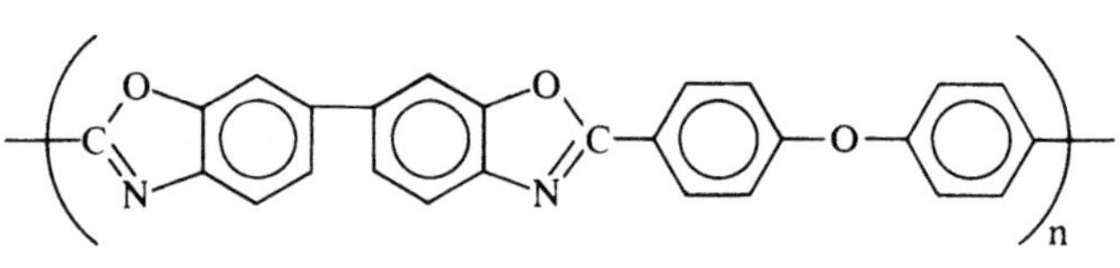

have been found to be photoconductive, with the photocurrent to dark current ratio of 5 x 10^2 – 5 x 10^3 at light intensity of 0.02 W/cm^2.

Dark currents in these polymers vary between 10^{-15} to 10^{-10} A/cm^2 in the fields up to 10^5 V/cm.

In general, the polymers with polyconjugated systems of double and triple bonds are photoconductive in the UV and at least part of the visible range. In some cases the photoresponse extends to the near infrared range. Although their usefulness in practical applications has been many times suggested, the results have been more or less disappointing. The main problems still remain: difficult synthesis, in most cases poorly identified structure, and with few exceptions insolubility and intractability of the polymers. The direct comparison with poly(N-vinyl carbazole) and other photoconductive polymers is not possible for lack of comparative data.

II. Polymers with Condensed Aromatic Groups

Photoconductivity has been detected in virtually all known polymers with condensed aromatic groups, either pendant to the main chain or within the main chain. The polymers with the simplest condensed aromatic nuclei, poly(1-vinyl naphthalene)[43)]

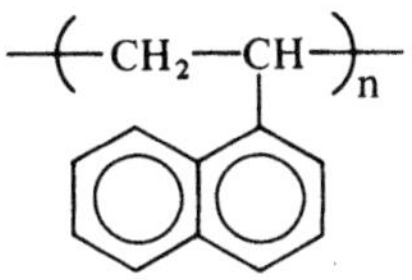

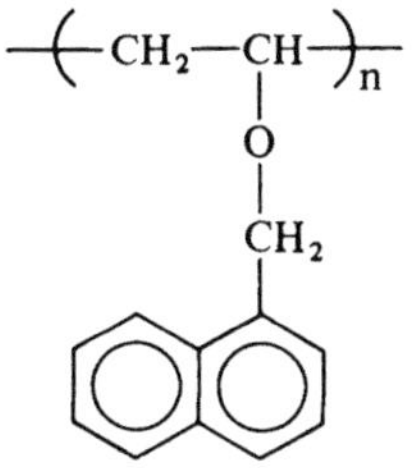

and poly(1-naphthyl methyl vinylether)[44)] showed only very small photocurrents in the UV range below 350 nm. The specific dark conductivities σ are near 10^{-18} 10^{-18} $ohm^{-1} cm^{-1}$. Poly(acenaphthene)[45)]

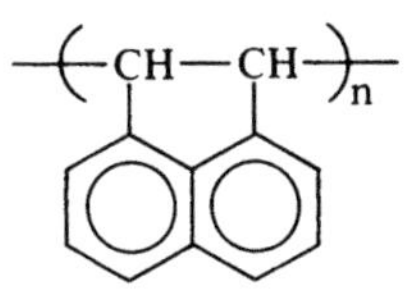

poly(vinyl acenaphthene)[46)]

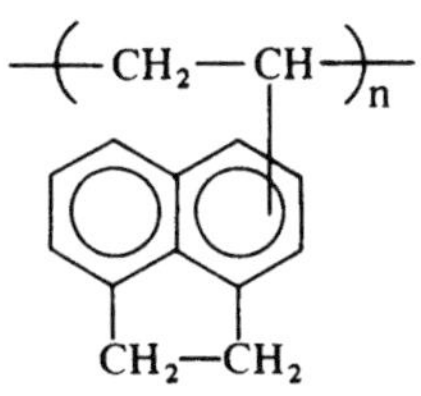

and poly(vinylacenaphthylene)[47]

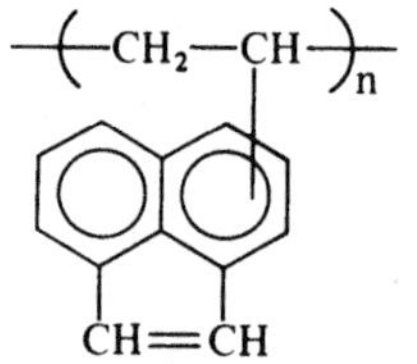

are also known as marginal photoconductors. The photo: dark current ratio in polyacenaphthene was only ~2 when using an incandescent lamp with 10^5 lux intensity. The dark current was near 10^{-11} A. Photoconductivity could be increased by partial nitration of the polymer.

Polymers with anthracene groups such as poly(9-vinylanthracene)[46,48], poly(1-vinylanthracene) and poly(2-vinylanthracene)[49,50], poly[1-(2′-anthryl)-ethylmethacrylate][51], poly(vinyl-p-phenyl-9-anthracene)[52–55], and poly(9-p-expoxy-

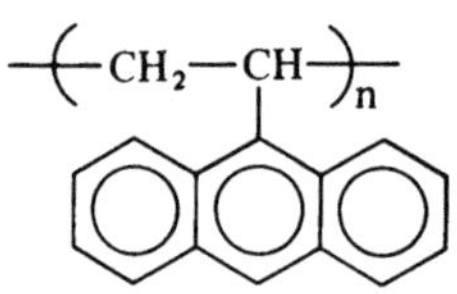

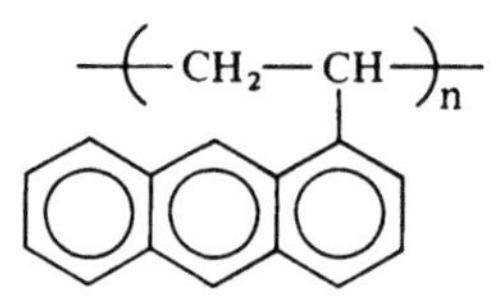

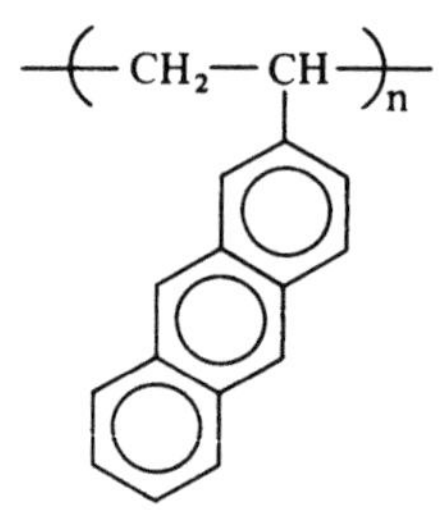

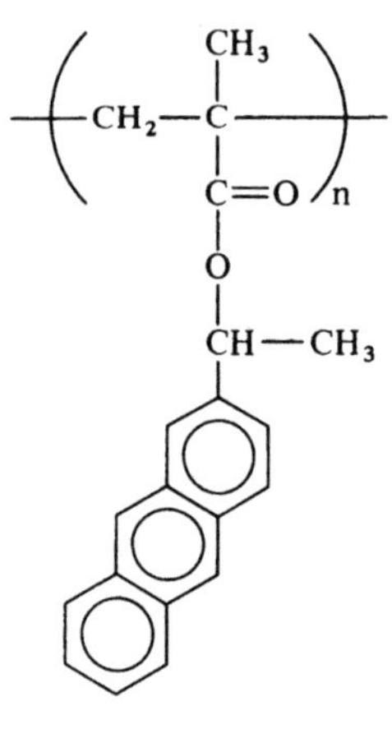

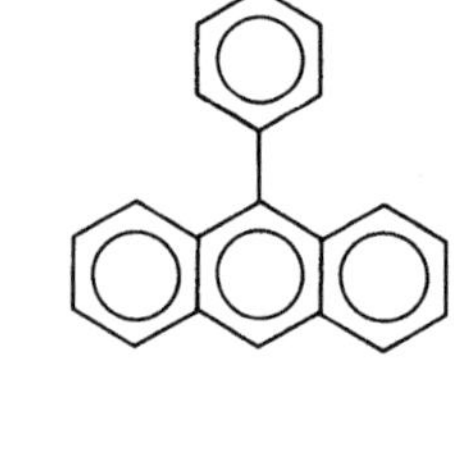

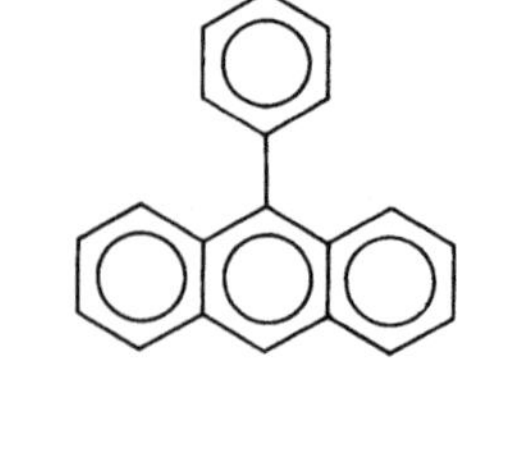

phenylanthracene)[56] are all photoconductive in the UV range of wavelengths. Perhaps with the exception of the last one, all anthracene polymers are brittle and poor film-formers. These polymers are also susceptible to oxidation and crosslinking by photodimerization of the anthracene groups in the UV light. The photoconductivity of the vinylphenylanthracene polymer is claimed to be comparable to that of PVK[54].

Poly(1-vinylpyrene)[46,57–59] has hole transport characteristics similar to PVK[59].

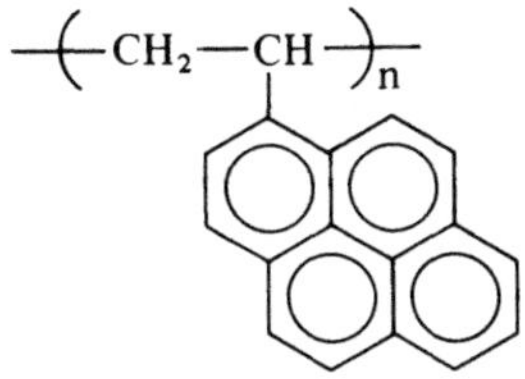

The hole mobility is 7.5×10^{-7} cm^2/volt sec. at a field of 2×10^5 V/cm. The mobility has an electric field dependence that varies between $E^{0.5}$ and $E^{1.5}$ in the field region of 2×10^4 V/cm and 2×10^5 V/cm[59]. An oligomer of 1-pyrenylmethylvinylether[44)

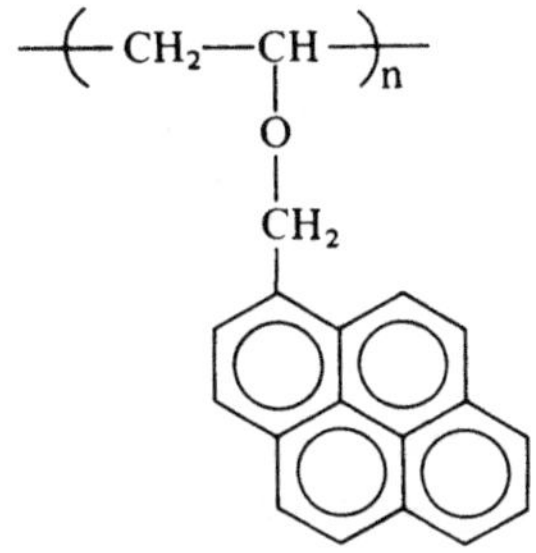

also exhibits substantial photoconductivity. The maximum photocurrent was observed at 410 nm while the absorption peak is at 376 nm.

The shift of the peak photoresponse towards longer wavelengths has been observed on some other polymers[60]. The magnitude of photocurrent of the TNF-sensitized poly(pyrenylmethylvinylether) is affected by the presence of oxygen[44]. A similar effect was observed on an undoped on an undoped polymer.

Another polymer with a large pendant polynuclear aromatic group is poly(9-vinylacridine)[61–63].

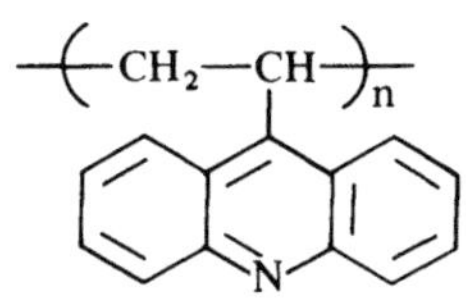

The polymer exhibits very low dark currents ($\sigma_{60°} = 3 \times 10^{-19}$ ohm^{-1} cm^{-1}). The photocurrent was reported to be proportional to the applied voltage and light intensity but its magnitude was far inferior to that of PVK. The poor photoconductivity is attributed to the high concentration of excimer forming sites acting as exciton traps and also to poor transport characteristics. The carrier transport is expected to be slower since the ionization potential of the polymer is higher (7.88 eV) than that of PVK (7.43 eV).

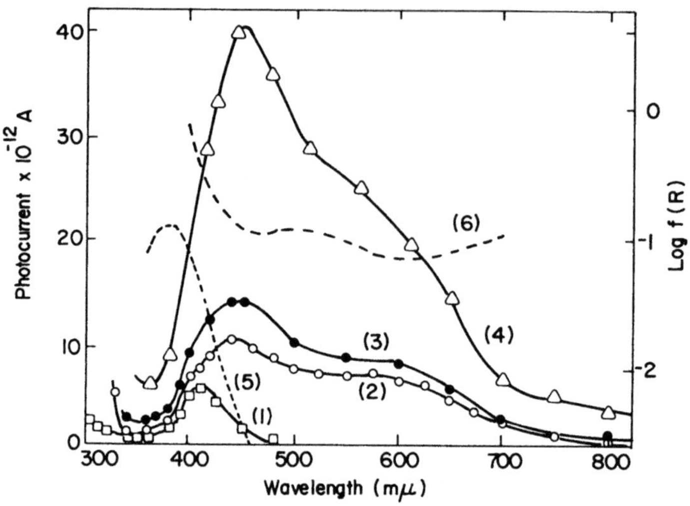

Fig. 8. Spectral responce of photocurrent of PPVE sensitized with TCNE and diffuse reflectance spectra both of PPVE and of PPVE doped with TCNE.

Photocurrent was normalized for the light intensity of 1×10^{14} photons/cm^2 sec. (1) PPVE in high vacuum, (2) PPVE doped with TCNE (1 mol%) in high vacuum, (3) PPVE doped with TCNE (1 mol%) in air, (4) PPVE doped with TCNE (10 mol%) in high vacuum, under 2500 V/cm. The values in the curve (4) are 1/2 of the observed ones. (5) and (6) diffuse reflectance spectra of PPVE and PPVE doped with TCNE (1 mol%)[44)]

Several other polymers with large pendant polynuclear aromatic groups have been recently synthesized[64)]. It is reasonable to expect that poly[3-(α acryloyloxy)-ethylperylene], poly(3-acrylamidoperylene) and poly(p-vinylbenzal-3-amidoperylene)[64)] will show photoconductivity and good carrier transport of carriers injected from selenium, charge transfer complexes, photoconductive dyes etc.

Photoconductivity has been observed on polymers to which the active polynuclear aromatic groups had been added by chemical attachment. For example, poly(L-lysine) with anthracene groups is claimed to be useful in electrophotography[65)]. Similarly, poly(p-alkyl glutamates) with the attached 9-anthryl, 9-anthrylmethyl and other groups were claimed to be photoconductive[66)].

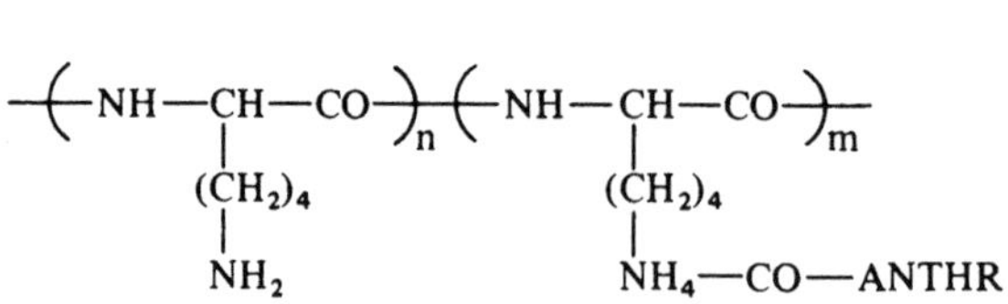

A number of photoconductuve resins made by condensation of anthracene, pyrene etc. with formaldehyde, benzaldehyde etc. have been disclosed in the patent literature. Their structure is however ill defined and no exact photoconductivity measurements have been reported in the literature.

The relatively high oxidative instability, difficulties associated with synthesis and generally poor mechanical properties have made most of the above polymers inattractive for practical utilization.

III. Polymers with Substituted Arylamino Groups

From the structural point of view, poly(N-vinylcarbazole) (PVK) is an example of many photoconductive polymers which have in common a substituted arylamino group:

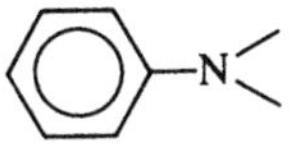

This group can appear in a great variety of configurations. The patent literature is particularly rich on photoconductive polymers which contain the arylamino group in all imaginable forms. Unless they contain an additional chromophoric group responding to visible radiation, these polymers are photoconductive only in the UV range of wavelengths. In the visible part of the spectrum they are generally nonabsorbing and act merely as charge transporting media. The charge carrier generators used in conjunction with the polymers of arylamine are either photoconductive dyes, various charge transfer complexes or inorganic photoconductors, such as selenium.

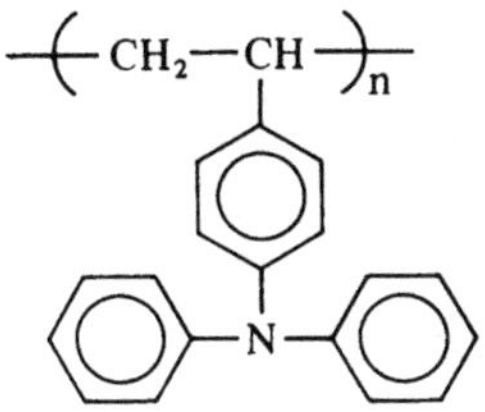

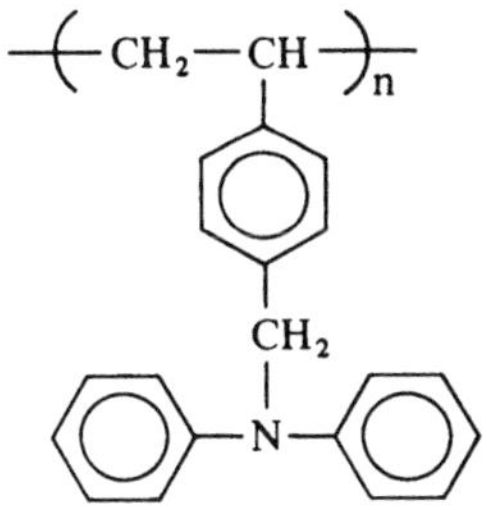

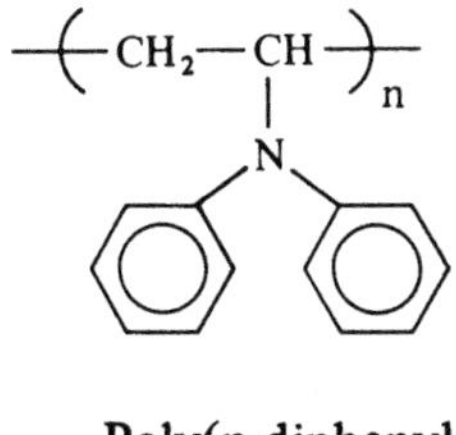

Poly(p-diphenylaminostyrene)[67], poly(N,N′-diphenylaminomethylstyrene)[68] and poly(vinyldiphenylamine)[69–71] are the simplest photoconductive polymers of this class. Similarly, photoconductive polymers claimed to be useful in electrophotography if properly "sensitized" were prepared from vinylesters and acetals containing triphenylamino groups[72]:

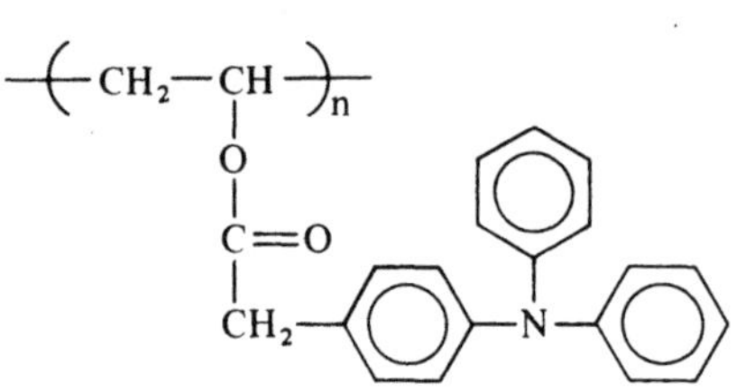

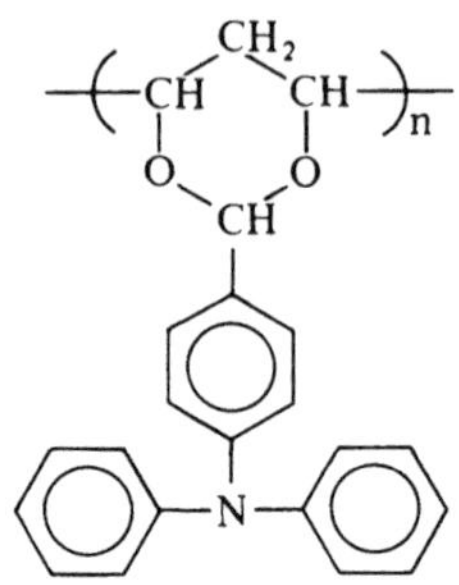

The patent literature disclosed several other arylamine containing photoconductive polymers such as polycondensates of diphenylbenzylamine with formaldehyde[73] poly(phenylbenzylamino-p-methylstyrene)[74] etc. Little can be said about these polymers apart from the fact that they exhibit photoconductivity. The carrier transport is undoubtedly provided by the arylamino group. The secondary factors such as the nature of the substituents, the relative overall concentration of the active groups, the distance from the polymer chain and the regularity of the structure will affect only the magnitude of the transport. Thus, very efficient hole transport was recently observed on four different arylamino containing methylmethacrylates[75]: poly[2-N-ethyl-N-m-tolylamino)ethylmethacrylate], poly(4-diphenylaminophenylmethyl methacrylate), poly[2-(N-methyl-N-phenyl)ethyl methacrylate], and poly(4-diethylaminophenylmethyl methacrylate). All four polymers are photoconductive in the UV range. The first two polymers have hole mobilities superior to

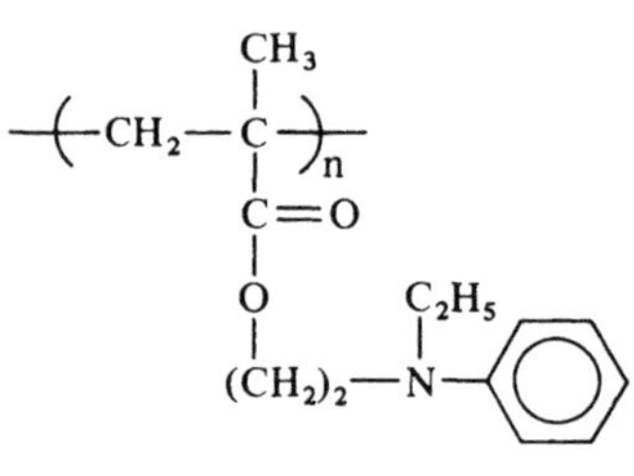

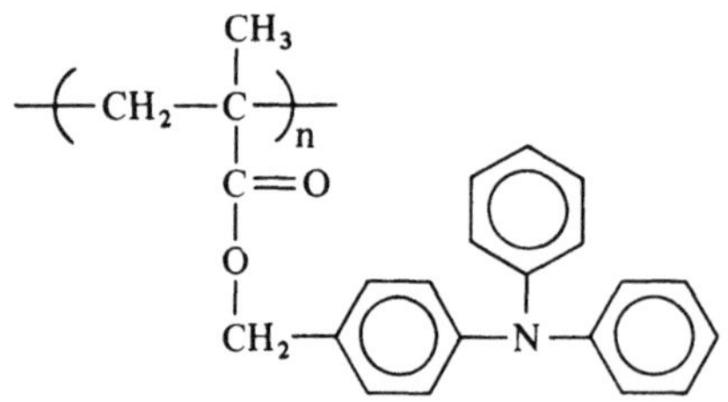

those of PVK. It can be seen from the data that the hole mobility in poly[2-(N-ethyl-N-m-tolylamino) ethylmethacrylate] is three times higher than the value measured in PVK. The electric field variation of the mobility is similar in both cases. The hole mobility in poly(4-diphenylaminophenylmethyl methacrylate) is 6×10^{-6} cm^2/volt

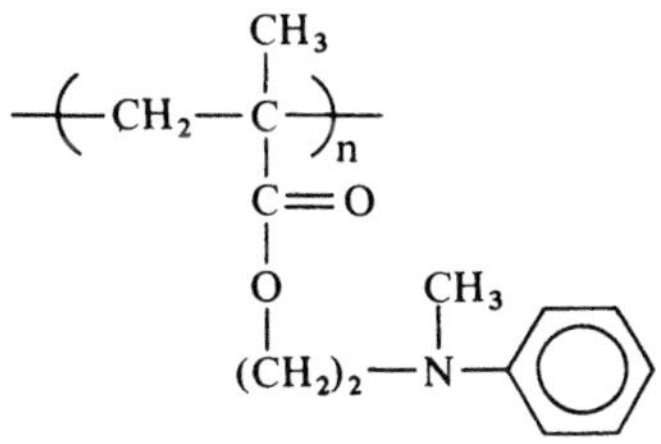

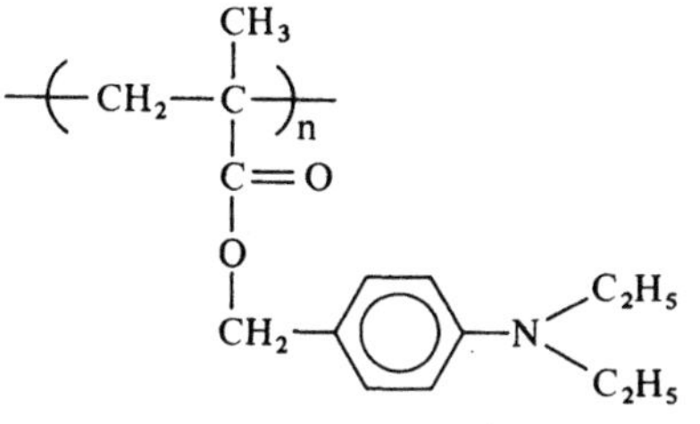

sec. at an applied electric field of 5 x 10^5 V/cm. This is about five times higher than the value of the mobility measured in PVK. The field dependence of mobilities in these polymers is similar to that in PVK[76].

Poly(N-vinylindole)[71,77] and its copolymers[78] have been reported as being useful in electrophotography when complexed with 2,4,7-trinitrofluorenone. Similarly, an acrylic polymer with indole groups is known to exhibit photoconductiv-

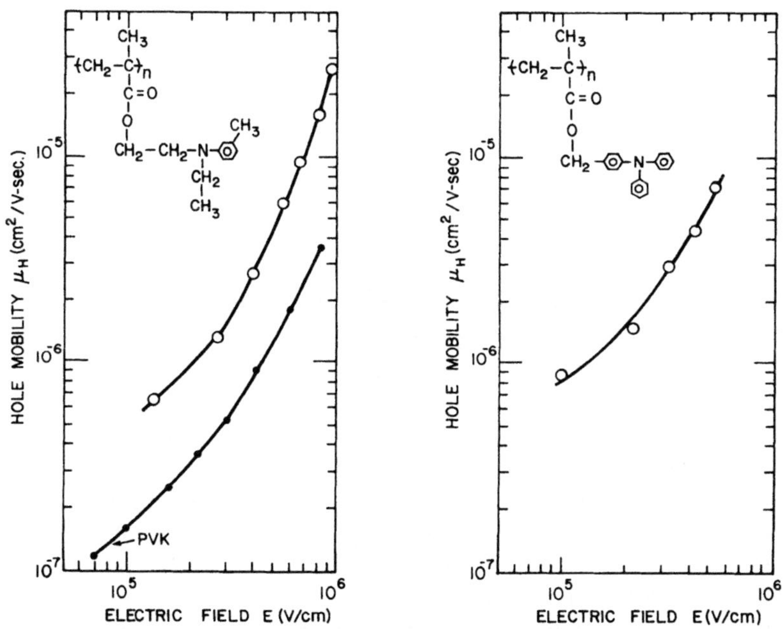

Fig. 9 (a). Hole mobility in a film of poly[2-(N-ethyl-N-m tolylamino) ethyl methacrylate] versus electric field is compared to that in PVK[75]. (b) Hole mobility in a film of poly[4-diphenylaminophenylmethylmethacrylate] versus electric field[75]

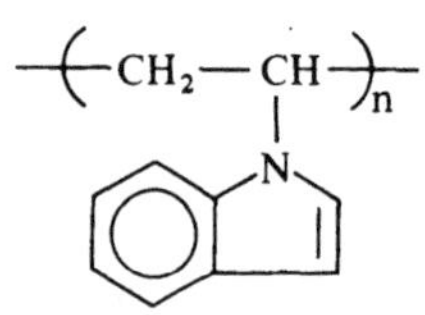

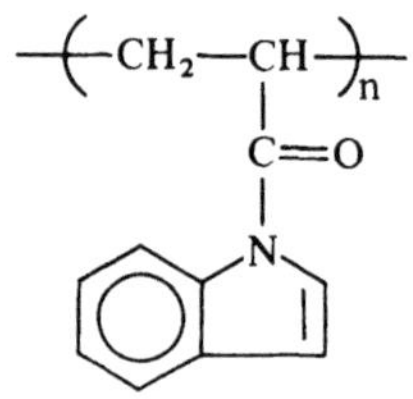

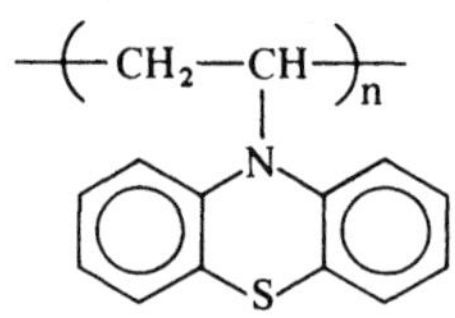

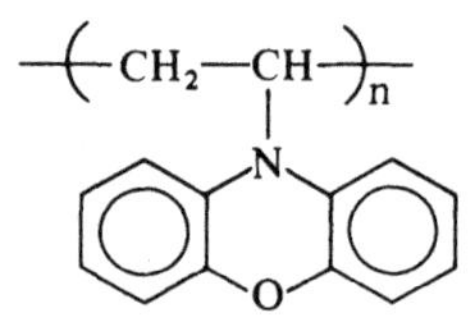

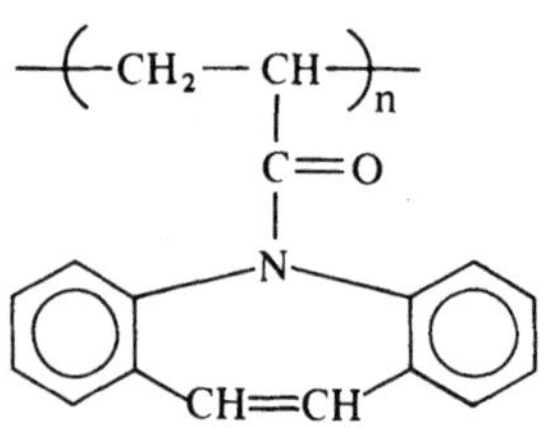

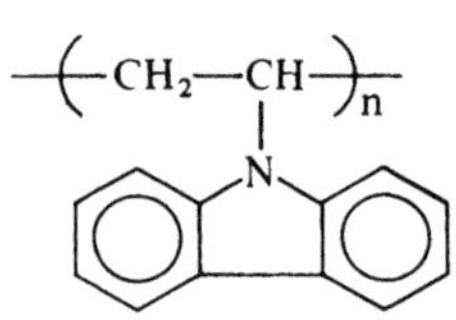

ity[79]. The following photoconductive polymers can also be classified as polymers of aromatic amines: poly(N-vinylphenothiazine)[71] and poly(N-vinylphenoxazine[70] and poly(N-acrylodibenzazepine)[80] Poly(N-vinylcarbazole)[81] is basically a modified vinyldiphenylamine polymer[69]. It has yet to be detemined if the transport characteristics of PVK with the diphenyl amino group forced into planarity are different from those of poly(N-vinyldiphenylamine) which would possess a greater freedom of rotation. The properties of PVK have been discussed in many articles and reviews [for example see Ref.[6]]. Several articles and patents have been published recently which deal with carbazole containing polymers other than PVK, and copolymers of N-vinylcarbazole with some other monomers.

The first group of these polymers contain structures in which a substituent was placed directly on the carbazyl ring. For example, bromine[82], iodine[83], nitroso[84], benzo[85], and iodobenzo[86] were added to the carbazole at various positions, apparently without a dramatic effect on the photoconductivity. A slight improvement of photoconductivity was claimed to be accomplished by placing an alkyl function in the 3-position of carbazole[87]. The second group contains those polymers in which the carbazyl group was separated from the backbone by one or more atoms. The highly crystalline homologues of PVK with alkylene groups between the polymer

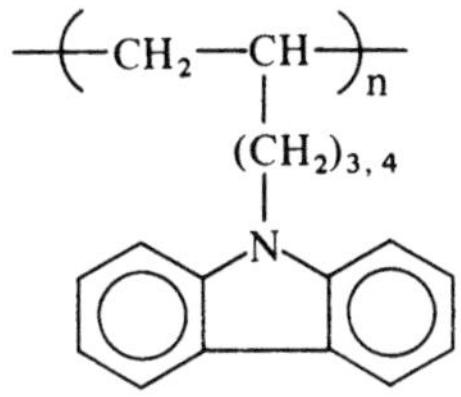

chain and the carbazole group were prepared by Heller *et al.* [88]. The photoconductivity of these polymers was established[89,90] but no direct comparison with PVK can be made at this time.

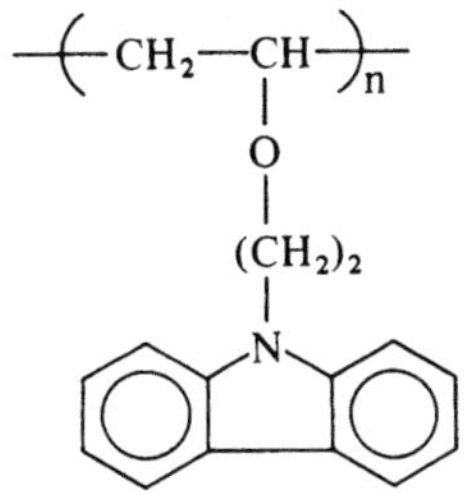

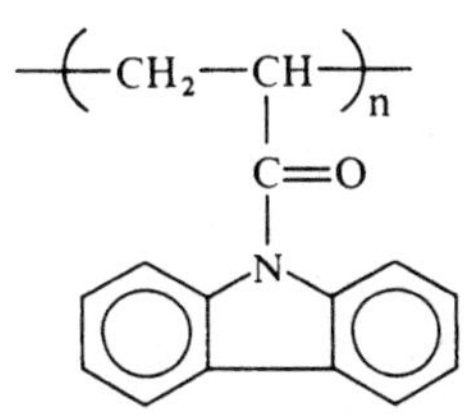

The photoconductivity of poly[2-(N-carbazolyl)ethyl vinylether][91,92] and particularly of poly(N-acryloylcarbazole)[91] is much inferior to that of PVK. In the case of the acrylic polymer the reported photocurrents are at least two orders of magnitude lower. The poor charge carrier generating efficiency is blamed for low photocurrents[91]. The relatively poor performance of the vinylether polymer is however attributed to charge carrier mobility[92].

Another photoconductive acrylic polymer, poly(N-acryloylaminopropylcarbazole)[93] is claimed to have good flexibility and adhesion properties.

Recently, Tanikawa *et al.*[94] discussed the photoconductivity of poly[γ-(β-N-carbazolyethyl)-L-glutamate] and its charge transfer complex with 2,4,7,-trinitro-9-fluorenone. The photocurrents in this polymer are about one order of magnitude smaller than those of PVK at all measured wavelengths. The complex with TNF has a peak photocurrent at ~600 nm while the absorption spectrum of the polymer

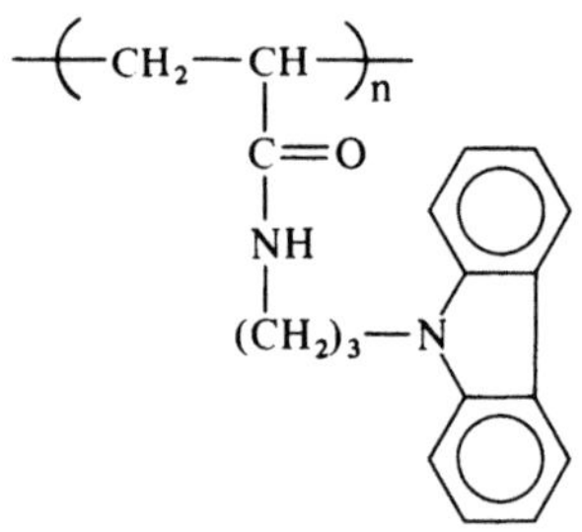

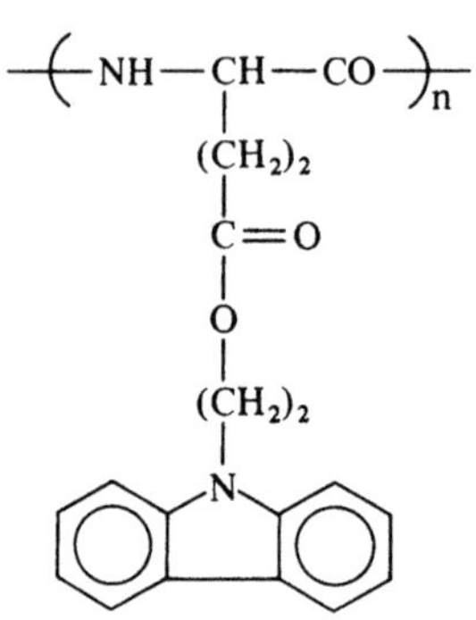

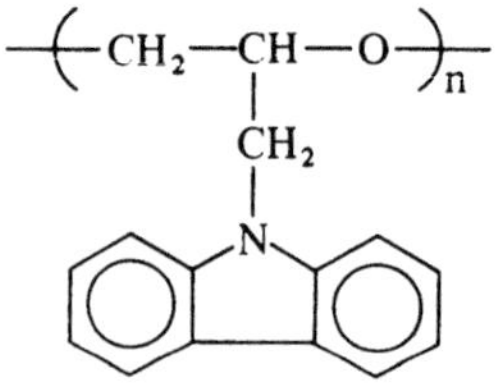

film has two shoulder peaks at ~460 and 540 nm. Gaidelis *et al.*[95] have reported drift mobility data on poly(N-epoxy propyl carbazole) having a molecular weight of between 1500 and 2000 and a glass transition temperature of approximately 50 °C. They found that the drift mobility is over an order of magnitude higher than the values reported for PVK[14]. The thermal activation energy of the drift mobility in this material was between 0.24 eV at 4.5×10^5 V/cm to 0.32 eV at 1.5×10^5 V/cm. These values are considerably lower than those reported for PVK[14]. The third group of carbazole polymers contain isomers of PVK: poly(N-ethyl-2-vinylcarbazole) and poly(N-ethyl-3-vinylcarbazole)[96,97]. The 2-isomer is reported to have higher drift mobilities than PVK while those of the 3-isomer are lower.

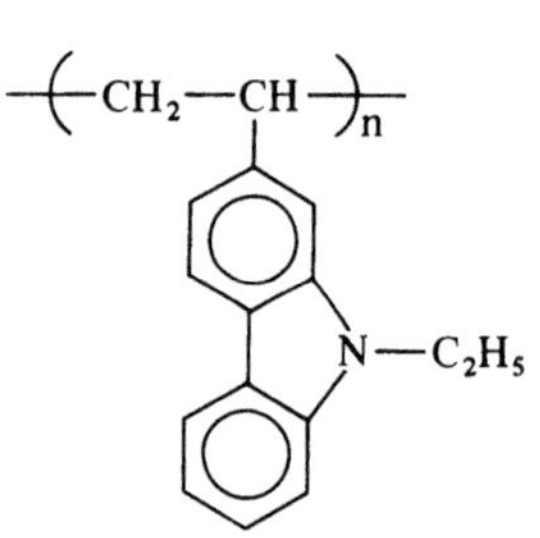

Copolymers of N-vinylcarbazole with styrene[98,99], alkyl and cycloalkyl methacrylates[98], vinylacetate and N-vinylpyrrolidone[99] and some other monomers[100] exhibit substantially lower photoconductivity than PVK. Only 10–12% of a comonomer such as styrene[101] is sufficient to cause a reduction of photoconductivity by one order of magnitude. Alternating copolymers of N-vinylcarbazole with fumaronitrile and diethylfumarate[99,101] exhibit no photoconductivity at all. The patent literature also describes a number of carbazole containing photoconductive resins, such as a polycondensate with formaldehyde[102]. Most of these resins are rather oligomers with poorly defined structures. Polymers of aromatic amines can assume other complicated forms. A polymer containing a pyrazoline ring[103] displayed

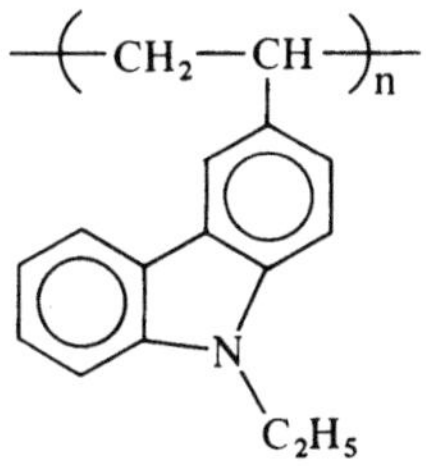

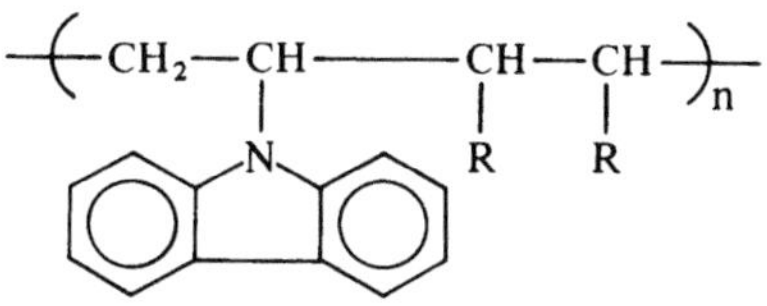

$R = CN; COOC_2H_5$

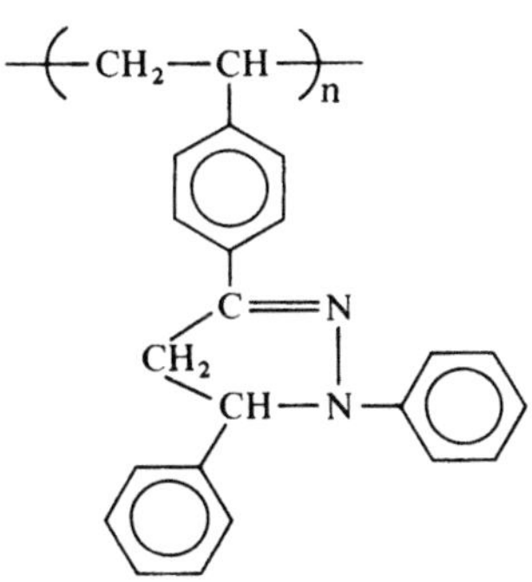

photoconductivity which was almost identical to that of 1,3,5 triphenylpyrazoline (model compound) mixed with an inert polyolefin resin. The maximum photoresponse occurred at 313 and 365 nm.

A derivative of indole, poly(6-vinylindole[2,3-b]quinoxaline)[104]

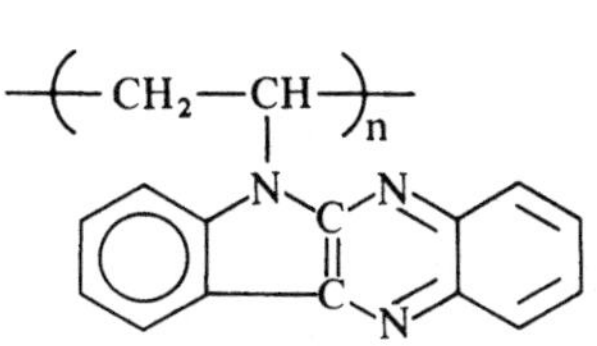

a polymer of the following structure[105],

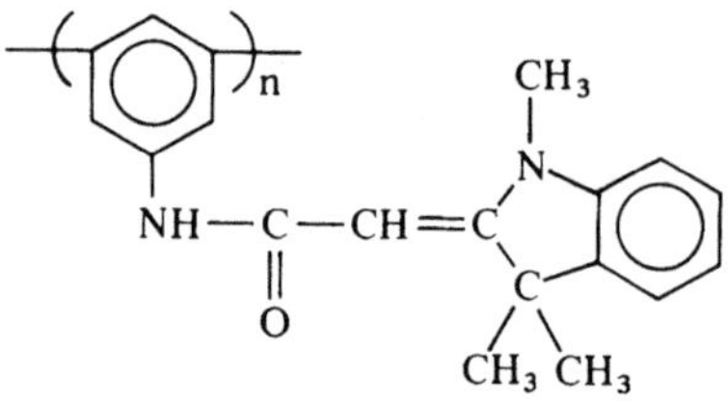

and a polycondensate of triphenylamine with sebacylchloride[106] are also reported to exhibit photoconductivity.

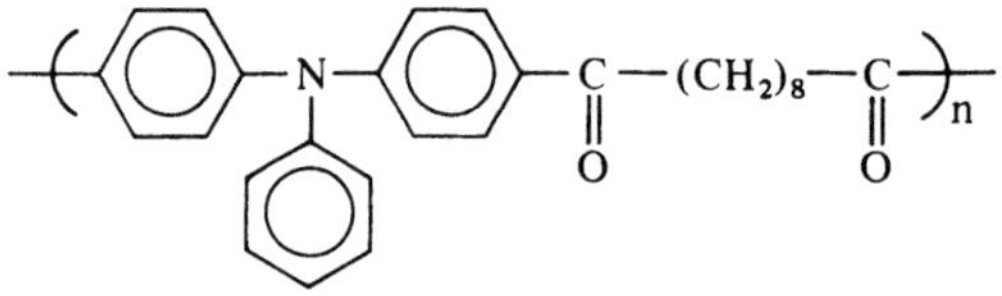

Many of the known arylamine containing polymers are potentially useful photoconductors. It is unfortunate that a direct comparison of most of these polymers with PVK is not yet possible. Each author used a different method of evaluation and the results are not comparable.

IV. Other Photoconductive Polymers

The literature refers to several other apparently marginally photoconductive polymers that do not fall into the above groups of polymers.

Gipstein and Hewett[107] synthesized the polymer and copolymer of 5-vinyl-2,2′-bithiophene which becomes photoconductive when complexed with large

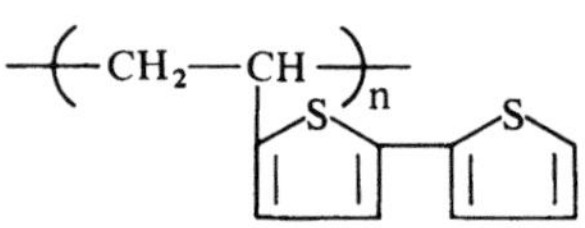

amounts of an acceptor such as TNF. Conceivably the carriers are generated via excitation of the bithiophene/TNF complex and the majority carriers are electron transported *via* TNF states.

Similarly, vinylferrocene copolymerized with a small amount of a film forming monomer displays photoconductivity[108] when complexed with 2,4-dinitrophenanthrolinequinone.

Some proteins have been reported as photoconductors. Eley and Metcalfe[109] measured nontransient photocurrents in hemoglobin, mitochondria and cytochrome C. The photocurrents were similar to those of DNA.

As mentioned above, most of the discussed photoconductive polymers are p-type photoconductors. Even in those cases where the charge carriers were not identified, it is safe to assume that the majority carriers are holes. All polynuclear aromatic and arylamine polymers are basically electron donors and therefore the transport involves equilibria between the neutral monomer units and positively charged units, cation radicals (holes). During the transport process the injected hole (cation radical) accepts an electron from the nearest neutral monomer unit in the direction of the electric field and thus the newly formed "hole" shifts towards the negative electrode. The process is repeated until the hole reaches the electrode.

Even in the first group of polyconjugated polymers, whenever the type of carriers was identified, it was always a positive type.

The only known electron transporting polymers known to date are polymers of a trinitrofluorenone monomer[110].

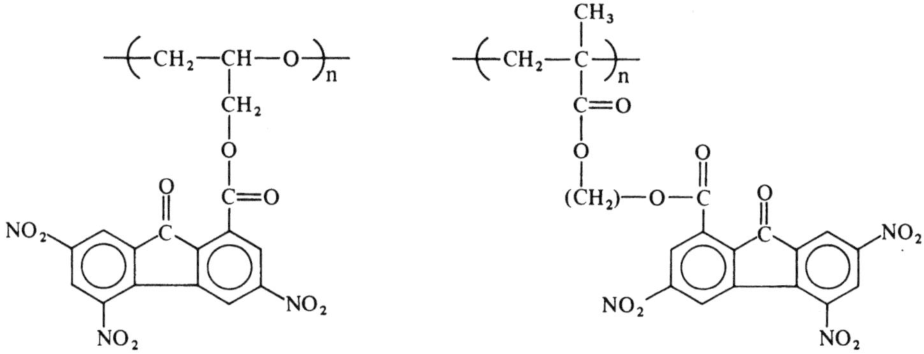

E. Structure-Property Relationships of Photoconductive Polymers

Some work was directed towards elucidation of the effect of microstructure on carrier transport properties of photoconductive polymers. Yoshimoto[111] and Williams[112] studied NMR spectra of poly(N-vinylcarbazole) and observed the large upfield shift of some aromatic protons. Williams[112] attributed the observed non-uniform shielding of the aromatic protons to the hindered rotation of the polymer segments and mutual interaction of the carbazyl groups. Limburg and Williams[96] proposed that a correlation may exist between the magnitude of the chemical shift and carrier mobilities in polymers. Such a correlation was observed on three vinyl-carbazole polymers[97], poly-(N-vinylcarbazole) (PVK), poly(N-ethyl-2-vinylcarbazole) (P2VK) and poly(N-ethyl-3-vinylcarbazole) (P3VK).

P2VK which showed the largest upfield shift of aromatic protons exhibited a hole mobility of 1.4×10^{-6} cm^2/V sec. at an electric field of 4×10^5 V/cm. PVK with slightly smaller shift had mobility of 1.4×10^{-7} cm^2/V sec. and P3VK with the smallest shift had hole mobility only 2.4×10^{-8} cm^2/V sec. at the same electric field. P2VK however exhibited the most severe charge carrier trapping.

Okamoto *et al.*[113] observed that the absorption spectra of vinyl polymers with large pendant π-electron systems including PVK show hypochromism a frequency shift in comparison with the spectra of the small molecule model compounds. He suggested that these polymers have a certain degree of order in the structure and proposed a direct relationship between the hypochromism and photoconductivity. On that basis, the poor photoconductivity of poly(β-N-carbazolylethylvinylether)[92], poly(acryloyl carbazole)[91], poly(1-vinylnaphthalene) and polyacenaphthylene[43]

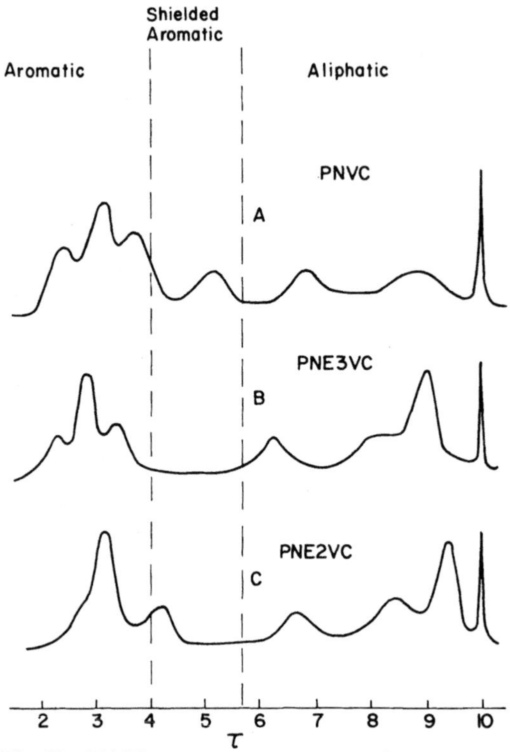

Fig. 10. 60-MHz proton nmr spectrum of PNVC, PNE3VC, and PNE2VC at room temperature in $CLCl_3$ [96)]

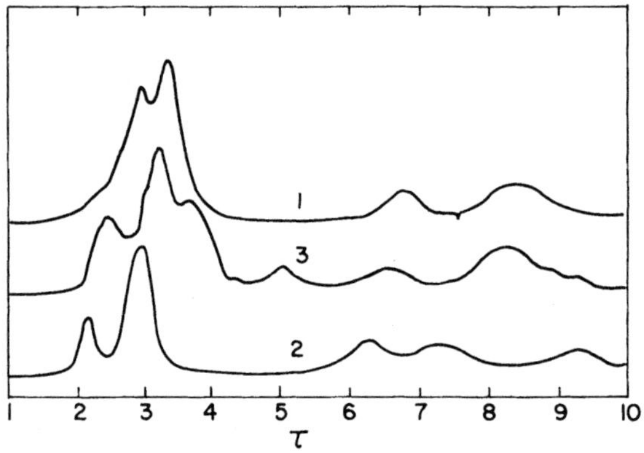

Fig. 11. 100-MHz NMR spectra of PACz, PCzEVE, and PVCz in $CDCL_3$. (1) PACz 60 °C; (2) PCzEVE, 20 °C; (3) PVCz, 80 °C [91)]

was explained by the lack of electronic interaction between the neighboring π-electron groups.

Okamoto also suggested that two indispensable conditions must be met for the polymer to be highly photoconductive[91)].

1) The pendant π-electron groups should be connected directly to the polymer chain,

2) the electronic interaction between the neighboring pendant π-electron groups should be strong enough to be clearly detectable by electronic and NMR spectra even in solution.

However, it is yet to be seen if this proposed direct correlation between the photoconductivity and nonuniform shielding of aromatic protons and the hypochromism of polymer spectra with respect to the monomer can be extended to other photoconductive polymers as well. The mutual interaction of the neighboring pendant groups may be a structural feature of polymers with bulky groups independent of charge transport. It is now known that some polymers with high carrier mobilities do not have bulky groups at all[75] and therefore do not satisfy at least the first condition of photoconductivity[92]. Poly(9-vinylacridine) on the other hand has rather poor photoconductivity mainly due to transport limitations[61] in spite of the close proximity of the group to the chain and clearly detectable interaction of the acridine groups[113]. Poly(9-vinylacridine) has of course relatively high ionization potential (7.88 eV vs. 7.43 eV for PVK)[114] which adversely affects the charge transport.

It is obvious that the limited data so far available on a very small number of photoconductive polymers do not permit definite conclusions of the role of group interactions.

It has also been suggested that the photoconductivity of PVK is associated with its crystallinity[115,116]. Poly(1-vinylpyrene) which is photoconductive[46,58,59] is also reported to have some degree of regularity[58]. A synthetic polypeptide, poly(γ-(β-N-carbazolylethyl)L-glutamate)[94] is proposed to have α-helix structure. The regularity of the chain conformation is also believed to lead to good photoconductive properties of the polymer.

F. Charge Carrier Generators

The above polymer were reviewed from the viewpoint of intrinsic photoconductivity, *i.e.* capability to both generate and transport charge carriers in the electric field. Polymers with polyconjugated systems of double bonds are intrinsic photoconductors not only in the UV region but most of them exhibit extended photoresponse into the visible range and some even to the near infrared region of the spectrum. The polymers with polynuclear aromatic groups and polymers with arylamino groups are intrinsic photoconductors only in the UV region. The photoconductivity of these polymers can be, however, extended into the visible range or occasionally even farther to the near infrared by appropriate "doping" or "sensitization". It is now commonly accepted that these "sensitizers" either constitute, or form in conjunction with the polymer, charge carrier generating species, capable of excitation, polarization and separation of the charge carriers in the electric field and injection of those carriers into the transporting polymer.

In accordance with the current views on the role of "dopants" or "spectral sensitizers" these charge carrier generating species can be formally divided into two groups: *Extrinsic, i.e.* those in which the polymer takes no part. These additives can either form homogeneous blends with the transporting polymer or form another phase or a layer. This category comprises for example vitreous selenium, various photoconductive pigments, dyes, etc. The second group are *complexing* additives, *i.e.* those which form a charge-transfer complex with the monomer units of the transporting polymer. The charge transfer complex then becomes the carrier generating species and can also contribute to the transport. The electron acceptor can also be chemically attached to the polymer. In such a case the polymer may become intrinsically photoconductive.

I. Extrinsic Photogenerators

Inorganic photoconductors such as selenium[5,76,117], selenium-tellurium alloys[118] and cadmium sulfide[119] are known to be capable of injection of photogenerated holes into PVK. Since the light penetration depth in selenium is relatively high (more than 90% of the incident radiation at 400 nm is absorbed within 0.2 nm from the surface), the carrier generation proceeds predominantly in the bulk of selenium[5] and only a very small portion occurs at the polymer interface. This in turn means that electron transfer (from the polymer) rather than energy transfer is involved in the carrier injection step. The general form of transit through PVK is the same as that which can be obtained by photoemission of hot carriers from a metal electrode or by intrinsic generation in PVK itself[117,120].

Selenium injects holes into other transporting polymers as well, *e.g.* poly(1-vinylpyrene)[59] and arylaminomethyl methacrylate polymers[75].

Regensburger[5] suggested that photosensitization by selenium involves hole generation and migration in selenium, injection across the interface and transport through the "sensitized" material. He also suggested that sensitization by dyes may involve the same type of mechanism.

Ikeda *et al.*[121,122] who studied sensitization of PVK by benzopyrylium and pyrylium salts provided evidence that the sensitization of the photocurrent in PVK and fluorescence quenching involve an electron transfer from PVK to an excited dye. The lower than expected quantum yield of photocarrier generation (<0.1)[122] is explained by the need for the thermal dissociation of the photogenerated charge transfer complex into free holes and electrons.

The photogeneration of carriers is field dependent[123] as it is in selenium[5]. This observation is in agreement with the previous suggestion[121] that the separation of positive-negative charge pairs in the excited dye is an intermediate step in charge carrier generation and injection. Hoegl[123] pointed to the fact that the injection of the hole is essentially oxidation of PVK and reduction of the dye.

The sensitization effect is not limited to the above mentioned dyes. For example, Terrel[98] studied charge carrier generation and injection processes in the crystal violet/PVK system. Meier *et al.*[124] extended the spectral response of PVK to over 600 nm by pinacyanol. Okamoto *et al.*[125] "sensitized" PVK into the visible region by Malachite Green, etc.

Dye sensitization of inorganic photoconductors was recently discussed by Tani[126] who also accepted the electron transfer mechanism of sensitization.

Table 2. Enhancement of Photoconductivity by Doping with Electron Acceptors[128)]

Dye	El. Acceptor	$\frac{I_{phot}\,(\text{doped})}{I_{phot}\,(\text{undoped})}$
Merocyanine A	O-Chloranil	60
	Iodine	1.7×10^3
Merocyanine FX79	p-Chloranil	6
	O-Chloranil	150
Phthalocyanine	Tetracyamethylene	400
	O-Chloranil	10^4

The patent literature discloses a large number of dyes as "spectral sensitizers" for various kinds of organic photoconductors including polymers. After reviewing a hundred or so patents dealing with this subject we can safely assume that all dyes which are known as photoconductive dyes[127–129)] will eventually inject carrier to transporting polymers with the efficiency varying from case to case. The field of application of the photoconductive dyes as spectral sensitizers is still virtually unexplored and direct structure-carrier generation and injection efficiency is still missing.

Meier[128)] observed that the photoconductivity of some dyes which are already known as efficient carrier generators can be enhanced by doping with electron acceptors (Table 2). He attempted to apply this effect to the PVK/dye/acceptor system[124)] using Methylene Blue and pinacyanol as carrier generators and tetracyanoethylene as an electron acceptor. Although no substantial increase in photocurrent was observed the spectral responce of the PVK/dye system was broadened to other wavelengths:

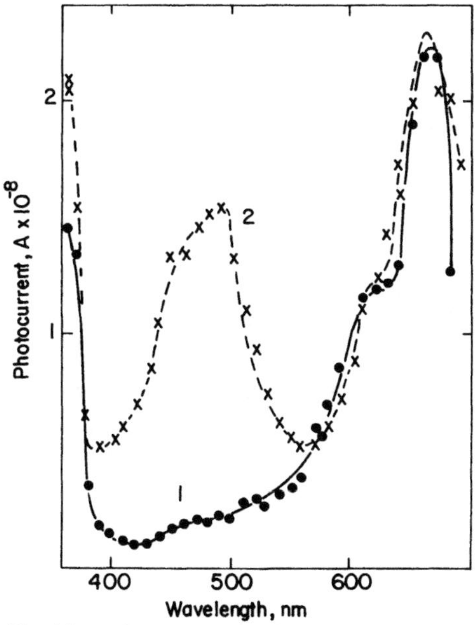

Fig. 12. Action spectrum of photoconductivity of sensitized PVK. (1) PVK + methylene blue. (2) PVK + methylene blue + TCNE[124)]

Besides the PVK band (in the UV range) and the dye band a new peak appeared in the photoconductivity spectrum. In the case of Methylene Blue it appears at 490 nm, *i.e.* between the polymer and the dye bands. This new peak is *not* a PVK/TCNE band since that appears near 600 nm and it is *not* a peak of the dye/TCNE complex since it does not vary with the nature of the dye. Although not supported by any other example in the literature, this can be an interesting concept of practical consequences.

Another group of extrinsic charge carrier generators are organic compounds containing both electron releasing and electron withdrawing groups such as N-ethyl-3-tricyanovinylcarbazol[130] or 1-dicyanovinylpyrene[131]. In these intramolecular com-

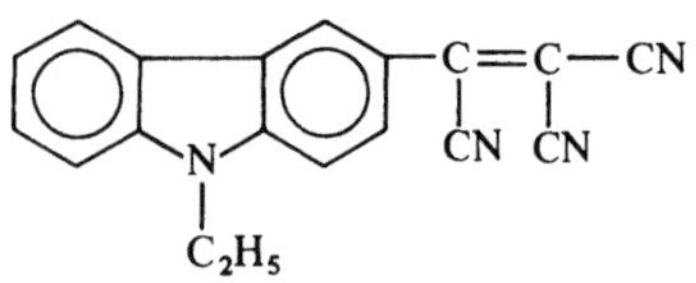

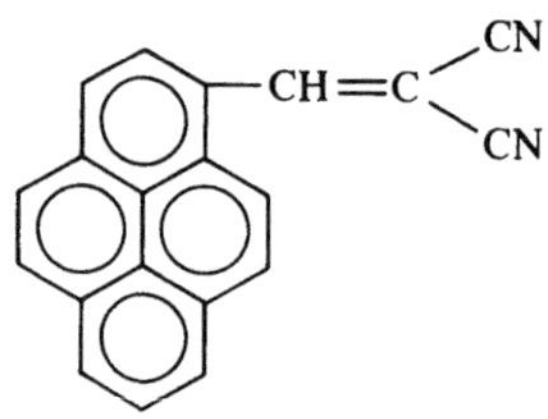

plexes the donor and acceptor groups are connected to one another through a spatially constraining linkage which insures that during photoexcitation of such a compound the electronic transition moment from ground to excited state and flow of charge between the groups are colinear. These compounds absorb light in the visible region and were used in conjunction with PVK and other transporting materials.

Dyes can be used as molecular dispersions (solid solutions) in the host polymer or as dispersions of crystalline dye particles/aggregates in the polymer matrix or eventually in the layered systems where the dye is dispersed in an inert polymer matrix such as polystyrene to form a thin charge carrier generating layer which is adjacent to the transport layer of PVK or another polymer[132]. In the first two cases the photogenerated holes are carried through the hole transporting matrix while electrons remain trapped in the dye unless an electron transporting medium is provided. In the third case the holes are moved through the polymer matrix towards the negative electrode (which can be a layer of corona-deposited charges) and the electrons are easily removed since they have to overcome only a short distance across the generator layer to the adjacent positive electrode. In this configuration it appears that the n-type photoconductive dyes could be more efficient than the p-type dyes.

II. Generation *via* Charge Transfer Complex Involving Polymer

The extension of sensitivity of photoconductive polymers from the UV to the visible range by doping with electron acceptors was reported by Hoegl[81,13]. Since all known photoconductive polymers have rather electron donating character and are hole transporting with the exception of polymers of 2,4,7-trinitrofluorenone[110)], the doping with electron acceptors results in formation of charge transfer complexes[134)]. The new spectral region of photosensitivity of the system corresponds to the absorption spectrum of the complex. Lardon, Lell-Doller and Weigl[134)] established a relationship between charge transfer excitation of the complex and the photoconductivity of the doped polymer. They proposed that the photoresponse of the "sensitized" polymer is accounted for by the photogeneration of carriers in the charge transfer complex and injection to the transport dielectric medium.

Among the electron acceptors which exhibit the "sensitization"[133)] are various nitro compounds, cyano-compounds, some organic acids such as phthalic acid, carboxylic acid anhydrides, quinones, etc. The most efficient electron accepting sensitizer known to date is 2,4,7-trinitro-9-fluorenone (TNF)[134,135)]. Not surprisingly,

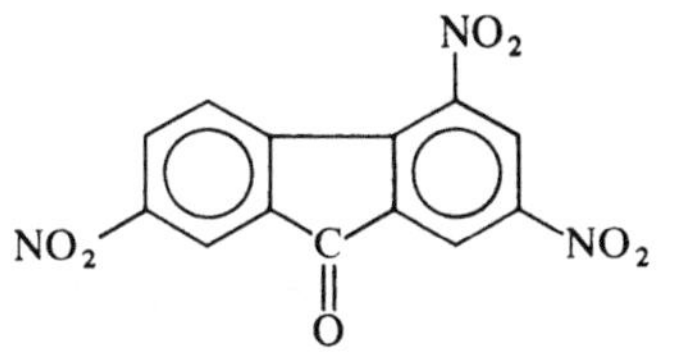

the PVK/TNF system was studied extensively by many authors[17,136–144)]. Schaffert[142)] described in detail the properties of this system and reported that the 1 : 1 molar mixture of PVK and TNF is almost panchromatic in the visible light. The white-light sensitivity is almost equal to that of amorphous selenium. The quantum efficiency is field dependent and reaches almost 0.5 at highest applied fields near 9×10^5 V/cm. Nominally the quantum efficiency is about 0.15 in the range of 400 to 500 nm and decreases slightly outside that region. In addition to the high sensitivity and high discharge rates the system has good dielectric strength, the 13 μ thick films could be charged to 1200 V. Gill found that hole mobilities of the PVK/TNF system decrease and the electron mobilities increase with increasing the ratio of TNF to PVK. A similar trend was observed by Tanikawa[94)] on a complex of poly[γ-β-N-carbazolylethyl)-L-glutamate] with TNF.

TNF alone, when dispersed in an inert polymer is a good electron transporting material[145,146)].

The generation and transport of charge carriers in these and other charge transfer complexes is determined by electron affinities and ionization potentials of the components involved and by the stabilization energies of cation radicals of the polymeric species and anion radicals of the acceptor, electron transporting species[147)].

The generation of charge carriers in charge transfer complexes is not limited to PVK/TNF systems. There have been several other studies reported involving other donor polymers and electron acceptors[44,85,86,89,93,133,148,149–152)] but none of the described system even approched the qualities of the PVK/TNF photoconductor. The quantum efficiency of generation is lower; for example, the quantum yield of carriers in PVK/tetracyanoethylene is in the order of 10^{-3} in the charge transfer band[14)] as compared with $\sim 10^{-1}$ of the PVK/TNF system[142)]. Poly(1-pyrenylmethylvinylether) alone is photoconductive in UV with peak response (~410 nm) which is slightly shifted towards longer wavelengths than the absorption peak (~390 nm). In the presence of TCNE the photoresponse shifts to longer wavelengths, well over 800 nm (Fig. 8). The shoulder in the photocurrent vs. wavelength curve is at 500–650 nm which is the charge transfer absorption region. The complexing with TCNE also enhanced the dark conductivity. Complexing of PVK with strong electron acceptors such as tetranitrofluorenone[138] or tetracyanoethylene[14] also resulted in substantially higher dark conductivity. Similar results were obtained with iodine[153]. Substantial dark currents were also achieved when poly(1-vinylpyrene) was complexed with tetracyanoethylene or, tetracyanoquinodimethane[150] or when polymers with phenothiazine units were complexed with tetranitrofluorenone[154]. In these cases the increase of dark currents went together with an increase of photoconductivity. The complexing of a strong donor polymer with a strong acceptor may lead to such an increase of dark conductivity[63,149,154–159] that the photocurrents may become insignificant.

It is interesting to note that doping of photoconductive polymers such as PVK with a strong electron donor such as tetramethyl-p-phenylene diamine reduced photoconductivity in all regions of wavelengths by a factor of up to 10^4 [14].

All the above examples involve charge transfer complexes between monomer units of the polymer and an *added* electron acceptor. There have been several cases reported in which both donor and acceptor groups become parts of the structure of the same polymer. For example, strong electron acceptor groups such as dinitrophenyl and chloranil were chemically attached to copolymers of N-vinylcarbazole with vinylacetate or acryloyl chloride[158]:

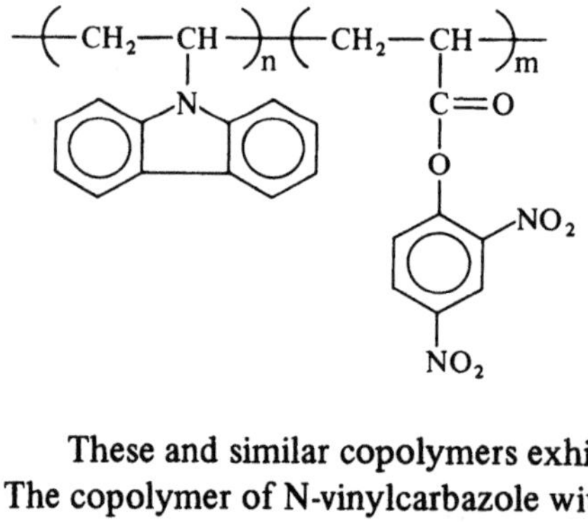

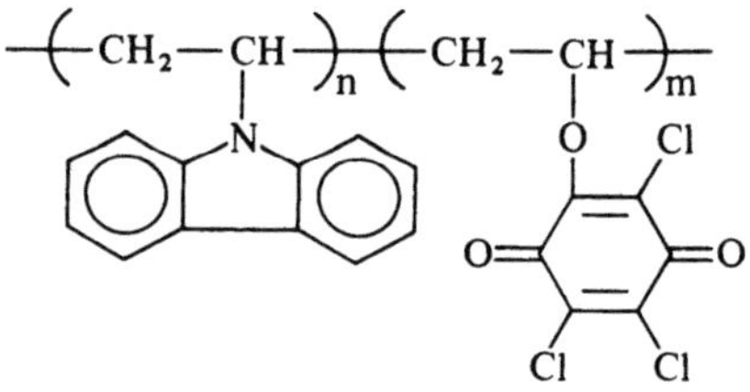

These and similar copolymers exhibited photoconductivity in the visible region. The copolymer of N-vinylcarbazole with dinitrobenzoyl groups showed better photoresponse than PVK doped with chloranil[158].

Photoconductivity of polyacenaphthylene was increased dramatically by partial nitration which produced acceptor groups in the polymer[45]. The naphthylene groups and the nitrated naphthylene groups apparently formed a charge carrier generating complex.

Photoconductive poly(pyromellitimides)[159] such as

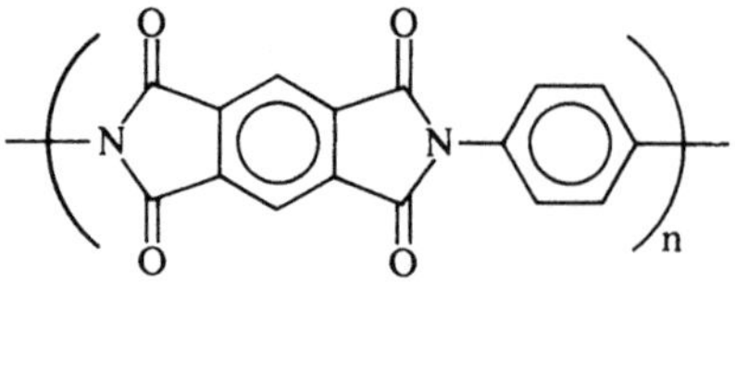

are basically poly-charge-transfer complexes between the arylamine electron releasing and pyromellitic electron withdrawing groups.

G. Conclusions

A large number of polymers with photoconductive properties has already been identified. However, with the exception of poly(N-vinylcarbazole) none of the photoconductive polymers has been studied in detail. In the case of poly(N-vinylcarbazole) and its charge transfer complex with 2,4,7-trinitro-9-fluorenone it has been clearly established that charge carriers can be transported through these materials without extensive deep trapping. This property, in addition, to the feasibility of preparation of large area, defect-free thin films by solvent casting were the main reasons for their successful application in electrophotography. The absence of deep traps in poly(N-vinylcarbazole) has made it possible to unambiguously determine the fundamental parameters such as charge carrier generation efficiencies, mobilities and recombination times. This is not the case with other photoconductive polymers. Only very few of the known materials can be compared with each other on the basis of published data. It is unfortunate since the existing photoconductive polymers cover a broad range of potentially useful physical and chemical properties. In order to obtain a meaningful comparison of various photoconductive polymers it is necessary to perform experiments which separate the carrier generation parameters from the parameters related to charge transport. Some of the published data suggest that the limiting factor is the carrier generation process and that systems involving the particular polymer would exhibit substantially better photoconductivity with an extrinsic carrier generator. Most of the published data do not permit the determination of the limiting step. Even in systems where the limiting step has been shown to be carrier trapping it is not clear if it is an inherent property of the material or an effect caused by extrinsic factors such as an accidental impurity, etc. From the data obtained so far on photoconductive polymers it has not been possible to establish criteria required for efficient transport. In spite of the obvious lack of fundamental information it is clear that the broad field of organic photoconductive polymer systems offers a great potential for the future.

H. References

1. Rose, A.: Concepts in Photoconductivity and Allied Problems. New York: Interscience Publishers 1967
2. Dolezalek, F. K.: Photoconductivity and Related Phenomena. Mort, J. and Pai, D. M. (ed.). New York: Elsevier Scientific Publishing Co. 1976, pp. 27–63

3. Batra, I. P., Kanazawa, K. K., Seki, H.: J. Appl. Phys. *41*, 3416 (1970)
4. Sharp, J. H.: Photoconductivity in Polymers. Patsis, A. V. and Seanor, D. A. (ed.). Westport: Technomic Publishing Co. 1976, pp. 98
5. Regensburger, P. J.: Photochem. and Photobiol. *8*, 429 (1968)
6. Pearson, J.: Pure and Appl. Chem. *49*, 463 (1977)
7. Pai, D. M.: Proc. Royal Electrophotographic Society Meeting, Cambridge, September 1976 (in press)
8. Onsager, L.: Phys. Rev. *54*, 554 (1938)
9. Geacintov, N. E., Pope, M.: Proc. Third International Conference on Photoconductivity, Pell, E. M. (ed.). Oxford: Pergamon 1971, pp. 289
10. Batt, R. H., Braun, C. L., Hornig, J. F.: J. Chem. Phys. *49*, 1967 (1968) and Appl. Opt. Suppl. *3*, 20 (1969)
11. Pai, D. M., Enck, R. C.: Phys. Rev. *11*, 5163 (1975)
12. Johnson, G. E.: J. Chem. Phys. *62*, 4697 (1975)
13. Yokoyama, M., Endo, Y., Mikawa, H.: Chem. Phys. Letters *34*, 597 (1975)
14. Okamoto, K., Kusabayashi, S., Mikawa, H.: Bull. Chem. Soc. Japan *46*, 2324 (1973)
15. Melz, P. J.: J. Chem. Phys. *57*, 1694 (1972)
16. Scher, H. and Montroll, E. W.: Phys. Rev. *B12*, 2455 (1975)
17. Gill, W. D.: J. Appl. Phys. *43*, 5032 (1972)
18. Weiser, G.: J. Appl. Phys. *43*, 5028 (1972)
19. Pai, D. M.: J. Appl. Phys. *46*, 5122 (1975)
20. Fox, S.: Photoconductivity in Polymers. Patsis, A. V. and Seanor, D. A. (ed.). Westport: Technomic Publishing Co. 1976, pp. 253
21. Pohl, H. A., Engelhardt, E. H.: J. Phys. Chem. *66*, 2085 (1962)
22. Mylnikov, V. S., Sladkov, A. M., Kudryavtsev, Yu. P., Lunieva, L. K., Korshak, V. V., Terenin, A. N.: Dokl. Akad. Nauk SSSR *144*, 840 (1962)
23. Mylnikov, V. S.: Uspekhi Khimii *43*, 1821 (1974)
24. Chance, R. R., Baughman, R. H.: J. Chem. Phys. *64*, 3889 (1976)
25. Schermann, W., Wegner, G.: Makrom. Chem. *175*, 667 (1974)
26. Horhold, H. H., Opfermann, J.: Makrom. Chem. *131*, 105 (1970)
27. Horhold, H. H.: Z. Chem. *12*, 41 (1972)
28. Horhold, H. H., Opfermann, J.: J. Prakt. Chem. *316*, 750 (1974)
29. Horhold, H. H., Graf, D., Opfermann, J.: Plaste u. Kautschuk *17*, 84 (1970)
30. Horhold, H. H., Opfermann, J.: Faserforschung u. Textiltechnik *25*, 108 (1974)
31. Horhold, H. H., Opfermann, J.: Makrom Chem. *178*, 195 (1977)
32. Mylnikov, V. S., Terenin, A. N.: J. Polymer Sci. *C16*, 3655 (1965)
33. Vasilenko, N. A., Pravednikov, A. N.: Vysokomolek. Soed., Ser. A *17*, 1741 (1975)
34. Cherkasov, Yu. A., Mylnikova, A. P., Polyakov, Yu. P.: Zh. Nauch. i Prikl. Phot.. Kinemat. *18*, 196 (1972)
35. Hirohashi, R., Hishiki, Y , Haruta, M.: Bull. Chem. Soc. Japan *44*, 2573 (1971)
36. Medvedeva, E. N., Kryazhev, Yu. G., Ermakova, T. G., Tatarova, L. A., Brodskaya, E. I., Poguda, T. S.: Vysokomolek. Soed., Ser. B *16*, 455 (1974)
37. Reucroft, P. J., Scott, H., Kronick, P. L., Serafin, F. L.: NASA Contract Rep. 1969, NASA-CR-1367; C. A. *71*, 61917s (1969)
38. Reucroft, P. J., Scott, H., Serafin, F. L.: J. Polymer Sci. C *30*, 261 (1970)
39. Bell, V. L., Pezdirtz, G. F.: Polymer Letters *3*, 977 (1965). Bell, V. L., Jewell, R. A.: J. Polymer Sci. A-1 *5*, 3043 (1967)
40. Voishchev, V. S., Rusanov, A. L., Leontieva, S. N., Kolnikov, O. V., Voishcheva, O. V., Mikhantiev, B. T.: Vysokomolek Soed., Ser. B *17*, 870 (1975)
41. Voishchev, V. S., Kolnikov, O. V., Kotov, B. V., Berendyaev, V. T., Voznesenskaya, N. N., Pravednikov, A. N., Sazhin, B. I.: Vysokomolek. Soed Kratkie Soobshcheniya *19*, 203 (1977)
42. Voishchev, V. S., Sazhin, B. I., Mikhantiev, B. I., Yakubovich, V. S., Gainulin, I. F.: Plast. Massy *1973*, 44
43. Morimoto, K., Murakami, Y., Ikeda, M.: Ref.[14] in 91
44. Yoshimoto, S., Okamoto, K., Hirata, H., Kusabayashi, S., Mikawa, H.: Bull. Chem. Soc. Japan *46*, 358 (1973)

45. Inami, A., Morimoto, K., Hayashi, Y.: Bull. Chem. Soc. Japan *37*, 842 (1964)
46. Hoegl, H.: Ger. Pat. 133,976 (1962)
47. Murakami, Y., Morimoto, K.: Brit. Pat. 1,125,548 (1968)
48. Morimoto, K., Ishida, E., Inami, A.: J. Polym. Sci. A-1 *5*, 1699 (1967)
49. Stolka, M., Yanus, J. F., Pearson, J. M.: Macromolecules *9*, 710 (1976)
50. Stolka, M., Yanus, J. F., Pearson, J. M.: Macromolecules *9*, 715 (1976)
51. Stolka, M.: Macromolecules *8*, 8 (1975)
52. Nozaki, S., Hagitani, A., Mukoh, A., Mori, Y., Sakashita, K.: Ger. Pat. 2,117,058 (1972); CA 77, 49151d (1972)
53. Nozaki, S., Hagitani, A., Mukoh, A., Mori, Y.: Ger. Pat. 2,146,104 (1972); CA *77*, 115076m (1972)
54. Mukoh, A., Mori, Y., Sakashita, K., Nozaki, S., Hagitani, A.: U.S. Pat. 3,764,590 (1973)
55. Hitachi Ltd.: Ger. Pat. 2,225,759 (1973)
56. Hitachi Seisakusho KK: Jap. Pat. 74,119,886 (1974)
57. O'Malley, J. J., Yanus, J. F., Pearson, J. M.: Macromolecules *5*, 158 (1972)
58. Tanikawa, K., Kusabayashi, S., Mikawa, H.: Polymer Letters *6*, 275 (1968)
59. Pai, D. M., Radler, R.: unpublished results
60. Okamoto, K., Kusabayashi, S., Mikawa, H.: Kogyo Kagaku Zasshi *73*, 1351 (1970)
61. Moriwaki, S., Okamoto, K., Kusabayashi, S., Mikawa, H.: Bull. Chem. Soc. Japan *48*, 2623 (1975)
62. Perrine, T. D.: J. Org. Chem. *25*, 1516 (1960)
63. Homer, R. B., Shinitzky, M.: Macromolecules *1*, 469 (1968)
64. Kamogawa, H.: J. Polymer Sci. A-1 *10*, 1345 (1972)
65. Nakanishi, T., Maruyama, K., Tashiro, T.: Jap. Pat. 74 05,470 (1974)
66. Hatano, M., Enomoto, T.: Jap. Pat. 74 75,688 (1972); CA *83*, 59940y (1975)
67. Fox, C. J.: U.S. Pat. 3,265,496 (1966)
68. Mitsubishi: Jap. Pat. 75,79638 (1975)
69. Hisatake, O., Honjo, S., Watana, O.: Ger. Pat. 2,007,962 (1970)
70. Sirotkina, E. E., *et al.*: Ger. Pat. 2,320,855 (1974)
71. Sirotkina, E. E., *et al.*: U.S. Pat. 4,038,468 (1977)
72. Merrill, S. H., Brantly, T. B.: U.S. Pat. 3,779,750 (1973)
73. Mitsubishi Paper Mills Ltd: Jap. Pat. 75, 1735150 (1975)
74. Mitsubishi Seishi: Jap. Pat. 74,102,345 (1974); CA *82*, 148519g (1975)
75. Stolka, M., Pai, D. M., Renfer, D. S., Yanus, J. F.: 175th ACS National Meeting, Anaheim, Ca., March 1978
76. Tabak, M. D., Pai, D. M., Scharfe, M. E.: J. Non-Cryst. Solids *6*, 357 (1971)
77. Gipstein, E., Hewett, W. A.: U.S. Pat. 3,554,741 (1971)
78. Gipstein, E., Hewett, W. A.: Ger. Pat. 1,917,747 (1969)
79. Hoegl, H., Schlesinger, H.: Ger. Pat. 1,131,988 (1962); U.S. Pat. 3,307,940 (1963)
80. Gipstein, E., Hewett, W. A.: Ger. Pat. 2,104,557 (1971)
81. Hoegl, H., Sus, O., Neugebauer, W.: Ger. Pat. 1,068,115 (1956)
82. Morimoto, K., Murakaim, Y.: Jap. Pat. 70 15,508 (1970); CA *73*, 56614c (1970)
83. Morimoto, K., Monobe, A.: Jap. Pat. 70 15,509 (1970), CA *73*, 56615d (1970)
84. Tubuko, K.: Ger. Pat. 1,772,036 (1972)
85. Dailey, E. A., Barton, J. M., Ginnis, R. A., Baltazzi, E. S.: Ger. Pat. 2,118,582 (1971)
86. Dailey, E. E., Barton, J. M.: U.S. Pat. 3,764,316 (1973)
87. Kobayashi, T., Suzuki, K., Murakami, H., Nishiide, K., Yamanouchi, T., Kinjo, K.: Jap. Pat. 73 59,843 (1973); CA *79*, 131380C (1973)
88. Heller, J., Lyman, D. J., Hewett, W. A.: Makrom. Chem. *73*, 48 (1964)
89. Hewett, W. A., Sporer, A. H.: U.S. Pat. 3,341,472 (1967)
90. Hewett, W. A.: U.S. Pat. 3,312,673 (1967)
91. Okamoto, K., Itaya, A., Kusabayashi, S.: Polymer *7*, 622 (1975); J. Polymer Sci., Polymer Phys. Ed. *14*, 869 (1976)
92. Turner, S. R., Pai, D. M.: Preprints, 175th ACS National Meeting, Anaheim, Ca., March 1978
93. Ito, H., Tazuke, S., Ohkavara, M.: Jap. Kokai 76,101,534 (1976), CA *86*, 16,3615 (1977)

94. Tanikawa, K., Okuno, Z., Iwaoka, T., Hatano, M.: Makrom. Chem. *178*, 1779 (1977)
95. Gaidelis, V., Krisciunas, V., Montrimas, E.: Thin Solid Films *38*, 9 (1976)
96. Limburg, W. W., Williams, D. J.: Macromolecules *6*, 787 (1973)
97. Williams, D. J., Limburg, W. W., Pearson, J. M., Goedde, A. O., Yanus, J. F.: J. Chem. Phys. *62*, 1501 (1975)
98. Terrell, D. R.: Photog. Sci. Eng. *21*, 66 (1977)
99. Okamoto, K., Kato, K., Murao, K., Kusabayashi, S., Mikawa, H.: Bull. Chem. Soc. Japan *46*, 2883 (1973)
100. Cassiers, P. M., Hart, R. M.: U.S. Pat. 3,155,503 (1964)
101. Shirota, Y., Yokoyama, M., Mikawa, H.: Polymer Preprints, ACS *14*, No. 1, 13 (1973)
102. Hirai, A.: Jap. Pat. *75*, 122,236 (1975)
103. Hirokazu, T., Takai, K., Kigoshi, F.: Chem. Abstr. *68*, 87787 (1968)
104. Ohta, M.: Jap. Pat. *75*, 89,399 (1975)
105. Ono, H., Osada, C., Watarai, S., Katsuyama, H.: Fr. Pat. 2,074,451 (1971)
106. Fox, C. J., Johnson, A. L.: U.S. Pat. 3,387,973 (1968)
107. Gipstein, E., Hewett, W. A.: U.S. Pat. 3,615,384 (1971)
108. Goto, T., Kojima, K.: Jap. Pat. *74* 93,018 (1974)
109. Eley, D. D., Metcalfe, E.: Nature *239*, 344 (1972)
110. Turner, S. R.: Ref.[38] in Pearson, J. M., Pure Appl. Chem. *49*, 463 (1977)
111. Yoshimoto, S., Akana, Y., Kimura, A., Hirata, H., Kusabayashi, S., Mikawa, H.: Chem. Commun. *1969*, 987
112. Williams, D. J.: Macromolecules *3*, 602 (1970)
113. Okamoto, K., Itaya, A., Kusabayashi, S.: Letters *1974*, 1167 (1974)
114. Okamoto, K., Ozeki, M., Itaya, A., Kusabayashi, S., Mikawa, H.: Bull. Chem. Soc. Japan *48*, 1362 (1975)
115. Kimura, A., Yoshimoto, S., Akana, Y., Hirata, H., Kusabayashi, S., Mikawa, H., Kasai, N.: J. Polymer Sci A-2 *8*, 643 (1970)
116. Nagao, M., Hermann, A. M.: Polymer Letters *12*, 69 (1974)
117. Mort, J.: Phys. Review B *5*, 3329 (1972)
118. Nishizima, H., Takahashi, M., Ohira, M., Endo, K.: Ger. Pat. 2,338,248 (1974)
119. Sato, K., Ototake, M., Yoshizawa, M.: Jap. J. Appl. Phys. *16*, 987 (1977)
120. Mort, J., Lakatos, A. I.: J. Non-Cryst. Solids *4*, 129 (1970)
121. Ikeda, M., Murakami, Y., Morimoto, K., Sato, H.: Jap. J. Appl. Phys. *9*, 931 (1970)
122. Ikeda, M., Morimoto, K., Murakami, Y., Sato, H.: Jap. J. Appl. Phys. *8*, 759 (1969)
123. Hoegl, H., Barchietto, G., Tar, D.: Photochem. Photobiol. *16*, 335 (1972)
124. Meier, H., Albrecht, W., Tschirwitz, U.: Photochem. Photobiol. *16*, 353 (1972)
125. Okamoto, K., Hasegawa, Y., Kusabayashi, S., Mikawa, H.: Bull. Chem. Soc. Japan *41*, 2563 (1968)
126. Tani, T.: Photogr. Sci. and Eng. *17*, 11 (1973)
127. Meier, H.: Zeitschrift für Wissenschaftliche Photographie and Photochemie *53*, 1 (1958)
128. Meier, H.: Topics Curr. Chem. *61*, 85 (1976)
129. Meier, H.: Organic Semiconductors; Dark and Photoconductivity of Organic Solids. Weinheim: Verlag Chemie 1975
130. Williams, D. J., Pearson, J. M., Limburg, W. W.: U.S. Pat. 4,006,017 (1977)
131. Pearson, J. M., Williams, D. J., Limburg, W. W.: U.S. Pat. 4,046,563 (1977)
132. Ikeda, M., Sato, H., Torii, E., Morimoto, K., Hasegawa, Y.: Ger. Pat. 2,150,527 (1972); U.S. Pat. 3,813,242
133. Hoegl, H.: J. Phys. Chem. *69*, 755 (1965)
134. Lardon, M., Lell-Doller, E., Weigl, J. W.: Mol. Cryst. Liq. Cryst. *2*, 241 (1967)
135. Shattuck, M. D., Wahtra, U.: Rep. of South Africa Pat. 672,820 (1968)
136. Queener, C. A.: Photogr. Sci. Eng. *15*, 423 (1971)
137. Hughes, R. C.: Appl. Phys. Letters *21*, 196 (1972)
138. Chiu, T. T., Hecht, J. K.: Tappi *54*, 391 (1971)
139. Hughes, R. C.: J. Chem. Phys. *58*, 2212 (1973)
140. Lakatos, A. I.: J. Appl. Phys. *46*, 1744 (1975)

141. Kanazawa, K. K., Sechtman, B. H.; Electrets, Charge Storage Transp. Dielectrics [2nd Int. Conf. 1972]; (publ. 1973), pp. 405-17. Perlman, M. M. (ed.). Electrochem. Soc., Princeton, N.J.
142. Schaffert, R. M.: IBM J. Res. Dev. *15*, 75 (1971)
143. Klopffer, W.: J. Chem. Phys. *50*, 2337 (1969)
144. Curtin, D. J.: Tapi *60*, 87 (1977)
145. Cherry, A. J., Neiman, R. R., Shattuck, M. U., Weiche, W. J.: Fr. Pat. 2,090,830 (1972); U.S. Pat. 3,791,826
146. Emerald, R. L., Mort, J.: J. Appl. Phys. *45*, 3943 (1974)
147. Rochlitz, J.: Chemiker-Ztg. *96*, 561 (1972)
148. Contois, L. E., Merrill, S., Grau, G. S.: U.S. Pat. 3,655,378 (1972)
149. Summers, J. W., Litt, M. H.: Electr. Prop. Polymers *1972*, 186. Frisch, K. C., (ed.): Tecnomic Publ., Westport, Conn.
150. Tanikawa, K., Ishizuka, T., Suzuki, K., Kusabayashi, S., Mikawa, H.: Bull. Chem. Soc. Japan *41*, 2719 (1968)
151. Hatano, M.: Jap. Pat. 75 15,515 (1975)
152. Ohta, M.: Jap. Pat. 75 100,067 (1975)
153. Hermann, A. M., Rembaum, A.: Polymer Letters *5*, 445 (1967)
154. Summers, J. W., Litt, M. H.: J. Polymer Sci. (Polym. Chem. Ed.) *11*, 1379 (1973)
155. Litt, M. H., Summers, J. N.: J. Polymer Sci. (Polym. Chem. Ed.) *11*, 1339 (1973)
156. Summers, J. W., Litt, M. H.: J. Polymer Sci. (Polym Chem. Ed.) *11*, 1353 (1973)
157. Litt, M. H., Summers, J. W.: J. Polymer Sci. (Polym. Chem. Ed.) *11*, 1359 (1973)
158. Chang, D. M., Gromelski, S., Rupp, R., Mulvaney, J. E.: J. Polymer Sci. (Polym. Chem. Ed.) *15*, 571 (1977)
159. Voishchev, V. S., Kolnikov, O. V., Gorbina, T. A., Kotov, B. V., Sazhin, B. I., Mikhantiev, B. I., Pravednikov, A. N.: Vysokomolek. Soed., Serv. B *16*, 295 (1974)

Received February 21, 1978
C. G. Overberger (editor)

Characterization of Graft Copolymers

Yoshito Ikada

Institute for Chemical Research, Kyoto University Uji, Kyoto, Japan

The product of graft copolymerizations and grafting reactions is mostly a mixture of A homopolymer, B homopolymer and A–B graft copolymer. The efficient methods for separating the graft copolymer from products and determining its purity are described. In addition, the chemical structure of separated graft copolymers, i.e., the number of branches, the length of branch and backbone polymers is discussed.

Table of Contents

I. Introduction

The graft copolymer is defined as a comb-like branched macromolecule in which the grafted side-chains have a chemical composition different from that of the backbone chain. The copolymer carrying only one grafted side-chain is also included in the classification of graft copolymers, unless the side-chain is attached to chain-ends of the backbone (block copolymer). Clearly, three characteristics are at least needed in specifying the molecular structure of a graft copolymer; the length of the backbone, the length of the side-chain and the number of the side-chains in one graft copolymer molecule. Other, less important specifications are the length distribution, *i.e.*, the molar mass distribution of both the backbone and the side-chain, the number distribution of the side-chain among the graft copolymer molecules, the spacial distribution of the side-chains on each backbone and the sort of chemical link at the branching site.

So long as the graft copolymer is not synthesized in any specially controlled fashion, it is awfully difficult to determine experimentally all these structural parameters. Even the average number of branches in one graft copolymer cannot be determined with good accuracy. Whether the characterization of graft copolymers is tedious work or not is markedly governed by the method adopted for the synthesis. If a graft copolymer is prepared by the radical polymerization of a monomer with the use of a radical initiator such as benzoyl peroxide, the yield of the graft copolymer would be fairly low, accompanied by homopolymerization. On the other hand, the graft copolymer would be much more effectively produced, if a living anionic polymer is allowed to react with another polymer that is capable of coupling the living polymer to form a graft copolymer.

Hitherto it has been rather conventional in graft copolymerization that studies of the kinetics and the graft copolymers formed have been conducted without specifying the above-mentioned basic structural parameters. Many works have been reported which do not attempt to isolate the true graft copolymer from the grafting product. This may be due largely to the fact that the purification as well as the characterization of graft copolymers is not easy to facilitate using conventional analytical means. Take for example the graft copolymerization of styrene onto polyethylene. A large amount of ungrafted polystyrene, *i.e.*, styrene homopolymer, will be unavoidably formed in addition to the grafted polystyrene chains. In order to study physical properties of the polyethylenestyrene graft copolymer as well as the kinetics of the graft copolymerization, we first of all need to separate the graft copolymer from the grafting product. How can we achieve this? Polystyrene is soluble in most of the solvents used for polyethylene. It is not simple to separate the grafting product into three components, the polyethylene-styrene graft copolymer, the ethylene homopolymer (unreacted polyethylene) and the styrene homopolymer. Even if the pure graft copolymer is successfully obtained, we must then determine the length and the number of the polystyrene side-chains. Suitable chemical reactions are not available at the present time to cut off selectively the polystyrene chains from the polyethylene backbone, as the structure of the branching site is supposed to be

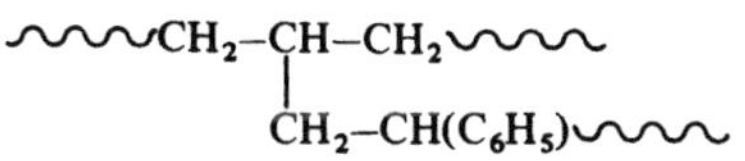

Although the purification and characterization of the graft copolymer are cumbersome, they must be made since they are needed in basic research as well as for the industrial applications of graft copolymers. In the following, we do not describe all the results published on the characterization of graft copolymers, but will discuss some fundamental problems relating to the characterization. Surprisingly few works have been published on this subject, though we have a tremendously large body of literature about grafting. Most of the results which will be shown in this review article have been accumulated in our laboratory.

II. Statistical Analysis of Graft Copolymerization and Grafting Reaction

As exemplified by the graft copolymerization of styrene onto polyethylene, the graft product often rejects even a simple characterization. In such a case, however, a statistical calculation, in other words, a simulation study of the graft copolymerization or the grafting reaction is helpful in understanding the molecular structure of the graft copolymer formed[1]. In addition such a calculation provides information on the yield of the graft copolymer as well as the amount of the unreacted mother polymer. To make the following statistical treatment applicable to both the graft copolymerization and the grafting coupling, we assume here that the grafted branches are generated randomly in the mother polymer and are independent of the branch length.

The basic equations of the grafting reaction are given by

$$\frac{\partial W_0(P,\alpha)}{\partial \alpha} = -PW_0(P,\alpha) \qquad \text{for } n = 0 \tag{1}$$

$$\frac{\partial W_n(P,\alpha)}{\partial \alpha} = -PW_n(P,\alpha) + PW_{n-1}(P,\alpha) \ \text{for } n \geqslant 1 \tag{2}$$

where $W_0(P, \alpha)$ and $W_n(P, \alpha)$ are the weight fraction of the mother polymer with the degree of polymerization P having no branch and n branches, resp., and α is the probability of a monomer unit to be effectively grafted in the mother polymer. For simplicity let us confine this calculation to the case that the molar mass of the mother polymer has a most probable distribution with $n \ll P$ and hence $\alpha \ll 1$. Then Eqs. (1) and (2) have the solution:

$$W_n(P,\alpha) = e^{-\alpha P} \frac{(\alpha P)^n}{n!} W(P,0) \quad \text{for } n \geqslant 0 \tag{3}$$

where W(P, 0) is the weight fraction at $\alpha = 0$, *i.e.*, the initial weight fraction of the mother polymer molecule with P.

1. Weight Fraction of the Mother Polymer with n Branches (W_n)

W_n is given by

$$W_n = \sum_{P=n}^{\infty} W_n(P, \alpha) = \sum_{P=n}^{\infty} e^{-\alpha P} \frac{(\alpha P)^n}{n!} W(P, 0) \quad (4)$$

In the case of the most probable distribution, $W(P, 0)$ is $[P/(\overline{P}_{B,O})^2]\exp(-P/\overline{P}_{B,O})$ where $\overline{P}_{B,O}$ is the number-average degree of polymerization of the mother polymer. Thus one obtains

$$W_n = (n + 1)(\alpha \cdot \overline{P}_{B,O})^n/(1 + \alpha \cdot \overline{P}_{B,O})^{n+2} \quad (5)$$

2. Weight Fraction of the Mother Polymer Grafted (w)

The weight fraction of the mother polymer possessing at least one branch is given by the following equation:

$$w = \sum_{n=1}^{\infty} W_n = 1 - W_0 \quad (6)$$

$$= \alpha \cdot \overline{P}_{B,O}(2 + \alpha \cdot \overline{P}_{B,O})/(1 + \alpha \cdot \overline{P}_{B,O})^2 \quad (7)$$

3. Average Number of Branches in One Mother Polymer Molecule (N_t)

In general N_t is given by

$$N_t = \sum_{n=1}^{\infty} n \cdot F_n \quad (8)$$

where F_n is the number fraction of the mother polymer with n branches. In the case of the most probable distribution F_n is written in the form

$$F_n = \sum_{P=n}^{\infty} \frac{i(P) \cdot {}_PC_n \cdot \alpha^n (1-\alpha)^{P-n}}{\sum_{P=1}^{\infty} i(P)} = \frac{1}{\overline{P}_{B,O}} \sum_{P=n}^{\infty} \exp(-P/\overline{P}_{B,O}) \cdot {}_PC_n \cdot \alpha^n (1-\alpha)^{P-n} \quad (9)$$

$$= (\alpha \cdot \overline{P}_{B,O})^n/(1 + \alpha \cdot \overline{P}_{B,O})^{n+1} \quad (10)$$

because i(P), the number of the mother polymer molecules with P, is

$$i(P) = (M/\overline{P}_{B,O})\exp(-P/\overline{P}_{B,O}) \quad (11)$$

where M is the total number of the mother polymer molecules. By substituting Eq. (10) into Eq. (8), we obtain the following equation:

$$N_t = \alpha \cdot \overline{P}_{B,O} \quad (12)$$

4. Average Number of Branches in One Graft Copolymer Molecule (N_g)
N_g is given by the following equation:

$$N_g = \sum_{n=1}^{\infty} n \cdot F_n / \sum_{n=1}^{\infty} F_n = 1 + N_t \tag{13}$$

It can be shown that N_g is related to w as follows:

$$N_g = -\ln(1 - w)/w \quad \text{(uniform distribution) and} \tag{14}$$

$$N_g = (1 - w)^{-1/2} \quad \text{(most probable distribution)} \tag{15}$$

5. Average Degrees of Polymerization of the Backbone with n Branches ($\overline{P}_{B,n}$ and $\overline{P}'_{B,n}$)

The number-average degree of polymerization of the backbone with n branches, $\overline{P}_{B,n}$, is given by

$$\overline{P}_{B,n} = \frac{\sum_{P=n}^{\infty} P \cdot i(P) \cdot {}_PC_n \cdot \alpha^n (1-\alpha)^{P-n}}{\sum_{P=n}^{\infty} i(P) \cdot {}_PC_n \cdot \alpha^n (1-\alpha)^{P-n}} \tag{16}$$

$$= [(n+1)/(1 + \alpha \cdot \overline{P}_{B,O})] \cdot \overline{P}_{B,O} \tag{17}$$

The weight-average degree of polymerization of the backbone with n branches, $\overline{P}'_{B,n}$, is

$$\overline{P}'_{B,n} = \frac{\sum_{P=n}^{\infty} P^2 \cdot i(P) \cdot {}_PC_n \cdot \alpha^n (1-\alpha)^{P-n}}{\sum_{P=n}^{\infty} P \cdot i(P) \cdot {}_PC_n \cdot \alpha^n (1-\alpha)^{P-n}} \tag{18}$$

$$= [(n+2)/(1 + \alpha \cdot \overline{P}_{B,O})] \cdot \overline{P}_{B,O} \tag{19}$$

6. Average Degrees of Polymerization of the Backbone of Graft Copolymer ($\overline{P}_{B,g}$ and $\overline{P}'_{B,g}$)

The number- and weight-average degrees of the polymerization of the whole backbone having any grafted branches, $\overline{P}_{B,g}$ and $\overline{P}'_{B,g}$, are given by the following equations:

$$\overline{P}_{B,g} = \frac{\sum_{n=1}^{\infty}\sum_{P=n}^{\infty} P \cdot i(P) \cdot {}_PC_n \cdot \alpha^n (1-\alpha)^{P-n}}{\sum_{n=1}^{\infty}\sum_{P=n}^{\infty} i(P) \cdot {}_PC_n \cdot \alpha^n (1-\alpha)^{P-n}} \tag{20}$$

$$= [(2 + \alpha \cdot \overline{P}_{B,O})/(1 + \alpha \cdot \overline{P}_{B,O})] \cdot \overline{P}_{B,O} \quad \text{and} \tag{21}$$

$$\overline{P}'_{B,g} = \frac{\sum_{n=1}^{\infty} \sum_{P=n}^{\infty} P^2 \cdot i(P) \cdot {}_PC_n \cdot \alpha^n (1-\alpha)^{P-n}}{\sum_{n=1}^{\infty} \sum_{P=n}^{\infty} P \cdot i(P) \cdot {}_PC_n \cdot \alpha^n (1-\alpha)^{P-n}} \tag{22}$$

$$= \frac{6 + 6\,\alpha \cdot \overline{P}_{B,O} + 2\,(\alpha \cdot \overline{P}_{B,O})^2}{(1 + \alpha \cdot \overline{P}_{B,O})\,(2 + \alpha \cdot \overline{P}_{B,O})} \cdot \overline{P}_{B,O} \tag{23}$$

7. Fractions of the Graft Copolymer with N Branches ($F_{g,n}$ and $W_{g,n}$)

If it is assumed that the formation of each branch is not affected by the position on the backbone polymer and the existence of other branches, the number and weight fractions of the graft copolymer with n branches are given by Eqs. (24) and (25), resp.:

$$F_{g,n} = \frac{\sum_{P=n}^{\infty} i(P) \cdot {}_PC_n \cdot \alpha^n (1-\alpha)^{P-n}}{\sum_{n=1}^{\infty} \sum_{P=n}^{\infty} i(P) \cdot {}_PC_n \cdot \alpha^n (1-\alpha)^{P-n}} \tag{24}$$

$$W_{g,n} = \frac{\sum_{P=n}^{\infty} (M_{O,B}\,P + nM_{O,b}\,\overline{P}_b) \cdot i(P) \cdot {}_PC_n \cdot \alpha^n (1-\alpha)^{P-n}}{\sum_{n=1}^{\infty} \sum_{P=n}^{\infty} (M_{O,B}\,P + nM_{O,b}\,\overline{P}_b) \cdot i(P) \cdot {}_PC_n \cdot \alpha^n (1-\alpha)^{P-n}} \tag{25}$$

where $\overline{P}_b$ is the number-average degree of polymerization of the branch polymer and $M_{O,B}$ and $M_{O,b}$ are the molar masses of the monomer residues of the mother and the branch polymers, respectively. When $M_{O,B} = M_{O,b}$ we obtain

$$F_{g,n} = (\alpha \cdot \overline{P}_{B,O})^{n-1}/(1 + \alpha \cdot \overline{P}_{B,O})^n \quad \text{and} \tag{26}$$

$$W_{g,n} = \frac{[n \cdot \overline{P}_b + (n+1)L](\alpha L)^{n-1}(1-\alpha L)^2}{\overline{P}_b + L(2-\alpha L)} \tag{27}$$

where $L = \overline{P}_{B,O}/(1 + \alpha \cdot \overline{P}_{B,O})$.

8. Number-Average Degree of Polymerization of the Graft Copolymer ($\overline{P}_g$)

Since the average number of branches in one graft copolymer is N_g, $\overline{P}_g$ is given by $(\overline{P}_{B,g} + N_g\overline{P}_b)$ and hence the following equation is obtained by using Eqs. (12), (13) and (21):

$$\overline{P}_g = \left(1 + \frac{1}{N_g} + N_g \cdot \frac{\overline{P}_b}{\overline{P}_{B,O}}\right) \overline{P}_{B,O} \tag{28}$$

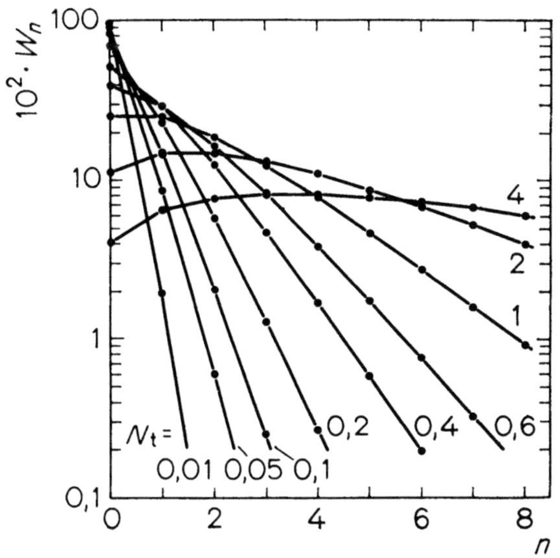

Fig. 1. Weight fraction W_n of backbone polymer with n branches for most probable molar mass distribution of the mother polymer; parameter is the average number of branches in one mother molecule, N_t. The points in the curves denote W_n at n = 0, 1, . . ., 8

We will graphically show the important results of the above calculation, some of which can be compared with experimentally obtainable data.

In Fig. 1 the weight fraction, W_n, of the backbone polymer with n branches is given as a function of the average number of branches in one mother polymer molecule N_t. As is clearly seen, the distributions of W_n become broader with increasing N_t. It is noteworthy that the weight fraction of the mother polymer with one branch (or two branches) does not exceed 0.2 (or 0.02), as long as N_t remains 0.01 to 0.1.

Figure 2 shows the relationship between the weight fraction of the grafted mother polymer w and the average number of branches N_g in one graft copolymer molecule. The plotting was performed according to Eqs. (14) and (15). The difference between the N_g–w relationship of the uniform and most probable distribution was negligible. The result denotes some important features in the grafting. For instance a graft copolymer with more than 2 branches on the average cannot be produced unless the weight fraction, w, of the mother polymer grafted becomes higher than 0.75 and 4.0% of the mother polymer remains ungrafted, even in the case of an average number of branches N_g = 5.

The relationship between $\overline{P}_{B,g}/\overline{P}_{B,O}$ and N_g is plotted in Fig. 3 according to Eq. (21). It can be seen that $\overline{P}_{B,g}/\overline{P}_{B,O}$ is always larger than unity and has a value of about 2 if N_g is close to unity.

In order to see the polydispersity of the molar mass of the backbone in the graft copolymer, the ratio $\overline{P}'_{B,g}/\overline{P}_{B,g}$ according to Eqs. (21) and (23) was plotted vs. N_g in Fig. 4. This ratio is seen to increase from 1.5 to 2 with increasing N_g, indicating that

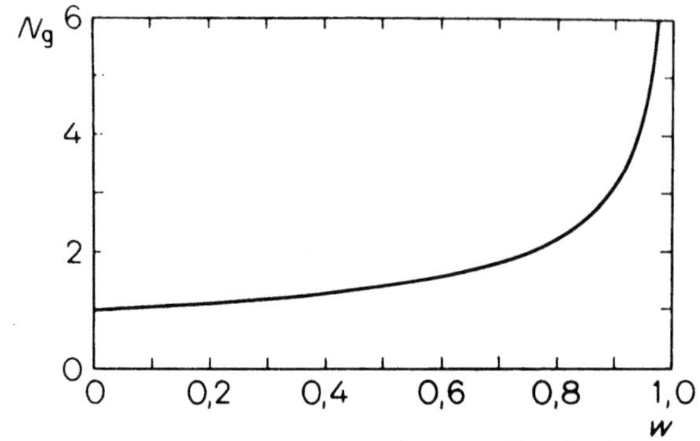

Fig. 2. Relation between weight fraction of grafted mother polymer, w, and average number of branches in one graft copolymer molecule, N_g

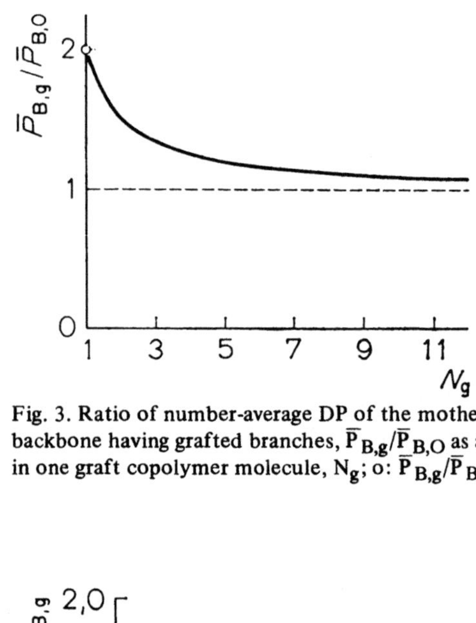

Fig. 3. Ratio of number-average DP of the mother polymer and number-average DP of the whole backbone having grafted branches, $\bar{P}_{B,g}/\bar{P}_{B,O}$ as a function of the average number of branches in one graft copolymer molecule, N_g; o: $\bar{P}_{B,g}/\bar{P}_{B,O}$ at $N_t \rightarrow 0$, *i.e.*, $N_g \rightarrow 1$

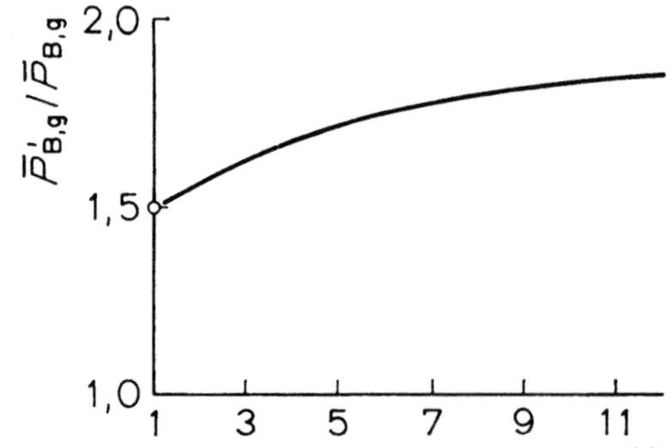

Fig. 4. Ratio of the weight- and number-average DP's of the whole backbone having grafted branches, $\bar{P}'_{B,g}/\bar{P}_{B,g}$ as a function of the average number of branches in one graft copolymer molecule, N_g; o: $\bar{P}'_{B,g}/\bar{P}_{B,g}$ at $N_t \rightarrow 0$, *i.e.*, $N_g \rightarrow 1$

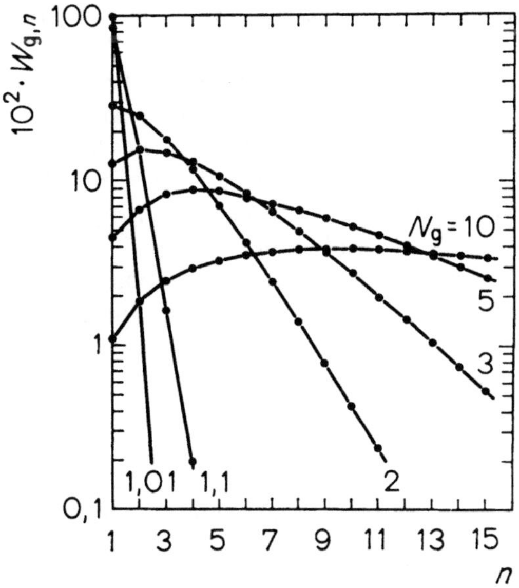

Fig. 5. Weight fraction $W_{g,n}$ of graft copolymer with n branches for the most probable molar mass distribution. Parameter is the average number of branches in one graft copolymer molecule, N_g. The points in the curves denote $W_{g,n}$ at n = 1, 2, . . ., 15

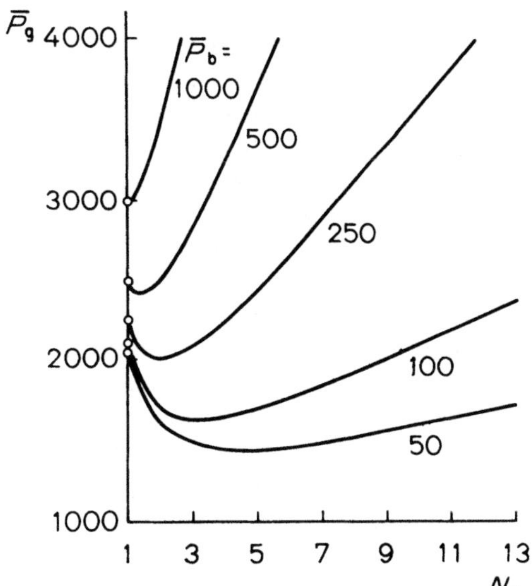

Fig. 6. Dependence of number-average DP of the graft copolymer, $\bar{P}_g$, on the average number of branches in one graft copolymer molecule, N_g. Number-average DP of the mother polymer, $\bar{P}_{B,O}$ = 1,000. Parameter is number-average DP of the branch polymer, $\bar{P}_b$; o: $\bar{P}_g$ at $N_t \to 0$, *i.e.*, $N_g \to 1$

the molar mass distribution of the backbone of the graft copolymer is more narrow than that of the mother polymer.

In Fig. 5 the weight fraction of the graft copolymer according to Eq. (27) was plotted vs. the number of branches for various N_g values. Here $\overline{P}_B$ and $\overline{P}_b$ were assumed to be 1,000. It is clear that most of the graft copolymer molecules have only one branch if N_g is close to unity. However, as the average number of branches increases, the grafted product should be regarded as a mixture of graft copolymers having branches whose number distribution is remarkably broad.

Finally we show in Fig. 6 the number-average degree of polymerization of the graft copolymer $\overline{P}_g$ as a function of the average number of branches N_g. The $\overline{P}_{B,O}$ in Eq. (28) was assumed to be 1,000.

The result shows the peculiar trend that with increasing N_g, $\overline{P}_g$ first decreases, if $\overline{P}_b$ is lower than $\overline{P}_{B,O}$. Clearly this is attributed to the polydispersity of the molar mass of the mother polymer (cf. Figs. 3 and 4).

As demonstrated above, we can learn much of the characteristics of graft products, from the statistical calculation, provided that the α value can be estimated either experimentally or theoretically. The α value in radiation graft copolymerization, for instance, can be evaluated from the G-value for radical production[2)].

III. Isolation of Graft Copolymers

The preceding calculation disclosed that a part of the mother polymer carries no grafted branch, even though the average number of branches in one graft copolymer molecule is as large as 5. Besides, it is often observed that the graft product contains the free homopolymer, which would otherwise constitute branches of the graft copolymer. If small amounts of homopolymers still contaminate the graft copolymer which has been isolated from the graft product, the molar mass of the graft copolymer determined, especially by the method which gives the number-average value, may be considerably low compared with the true molar mass of the graft copolymer, free of any homopolymers. Therefore a rigorous isolation of the graft copolymer from the graft product should be the first step in the characterization of the graft copolymer.

We will discuss several problems related to the isolation process based on our experimental results and present effective isolation methods, especially noting the emulsifying effect of graft copolymers on the isolation. For illustrative purposes attention will be confined to the graft copolymers with one branch, but similar considerations may apply to those with many branches and block copolymers.

The methods generally utilized to isolate the graft copolymer from the reaction product can be, in principle, divided into two groups; selective precipitation and extraction.

A. Selective Precipitation

Selective precipitation is usually carried out with the intention to precipitate only one homopolymer, say poly-A, from the solution of a crude reaction product by adding a precipitant for poly-A. The separation by this method seems to give successful results, since the poly-A-B graft copolymer can be held in the solution as dispersed micelle[3]. However, on addition of the precipitant a stable dispersion is usually formed which cannot be coagulated to a complete extent even by a long centrifugation at high gravities.

This phenomenon may be interpreted as follows: with the addition of the precipitant, not only the poly-A homopolymer but also the poly-A sequence in the graft copolymer molecule collapse and form the core of the micelle. The other, soluble sequences may form a peripheral outer shell of the micelle. In other words, the graft copolymer is considered to act as an emulsifier of poly-A homopolymer. This emulsification is schematically represented in Fig. 7. For instance, the homopolymer to be precipitated is completely emulsified by a graft copolymer of 1.5 wt.% of the homopolymer[4]. Accordingly, such a method intended to selectively precipitate only one homopolymer does not give the true purification because of the emulsifying ability, which is a characteristic of the coexisting graft copolymer.

However, if it is possible to coprecipitate the graft copolymer with poly-A, keeping poly-B alone in the solution, we may be able to remove poly-B from the reaction product. For this purpose the solubility of the poly-A-B graft copolymer should sufficiently differ from that of poly-B. As an example, turbidimetric titration curves of polystyrene(PS), poly(vinyl acetate)(PVAc), and the PVAc-styrene graft copolymer isolated from a radiation-grafting product are shown in Fig. 8 [5]. The turbidity was measured by adding water, a common non-solvent, to each dioxane solution. The result indicates clearly that the graft copolymer is completely coagulated before the PVAc homopolymer begins to be precipitated, suggesting that our proposed method is promising.

Table 1 gives a typical result of this modified precipitation performed for a physical mixture of PVAc, PS and PVAc-styrene graft copolymer (1 : 1 : 0.5). The

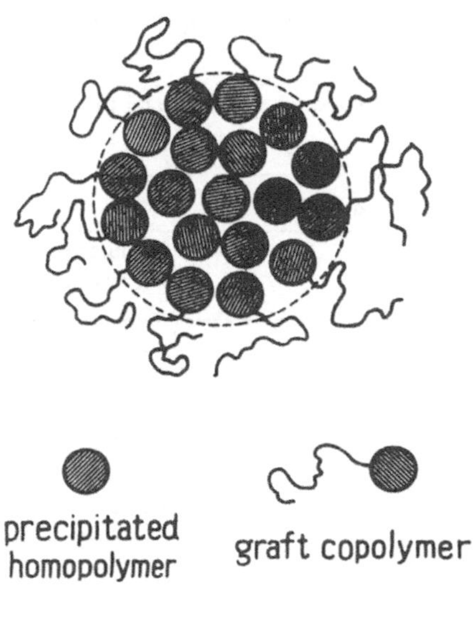

Fig. 7. Schematic representation of the emulsifying mechanism

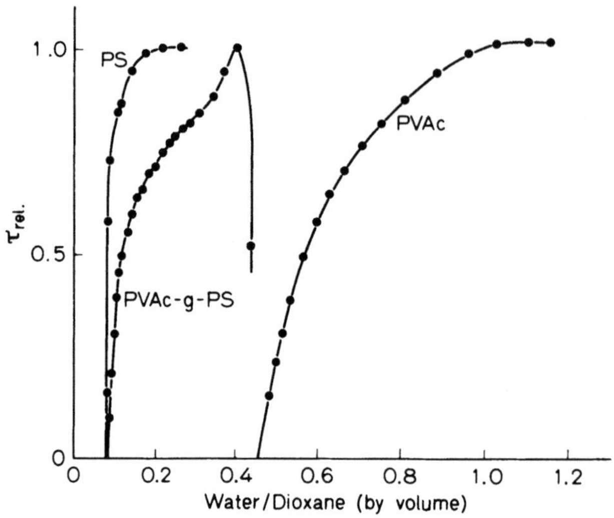

Fig. 8. Turbidimetric titration curves of PS, PVAc and PVAc-styrene graft copolymer (PVAc-g-PS)

mixture is dissolved in dioxane and precipitated selectively by adding water in such an amount that PS is completely precipitated but PVAc still remains in the solution. After the solution-precipitate mixture is subjected to centrifugation at a high speed, the PVAc homopolymer is recovered by evaporation of the solvent from the supernatant solution. As seen in Table 1, only about half of the added PVAc homopolymer is separated by the first precipitation procedure. However, the coprecipitated PVAc homopolymer can be removed by repetition of the re-solution and re-precipitation cycle for the precipitate. In this experiment the separation was accomplished after 6 repetitions. The number of repetitions required for the complete separation decreases if the precipitant is added in small amounts so as just to coagulate the graft copolymer and is added as slowly as possible. When the amount of the precipitant is too insufficient to coagulate the graft copolymer completely, the supernatant

Table 1. Removal of PVAc from a PVAc-PS-PVAc-styrene graft copolymer mixture by the modified precipitation (initial polymer conc. = 2 g/dl)

Precipitation	Water / Dioxane (by vol.)	Recovered PVAc Weight (mg)	Recovered PVAc Cumulative wt. fraction
1	0.35	51.4[a]	0.514
2	0.35	25.3[a]	0.767
3	0.365	3.9	0.806
4	0.35	18.5[a]	0.991
5	0.36	0.5	0.996
6	0.36	0.6	1.002

[a] Styrene unit was not detected in the IR spectra.

appears more or less turbid even after the centrifugation. The removal of the PS homopolymer from the PVAc-PS-PVAc-styrene graft copolymer mixture is also successful when dioxane-n-hexane is used as a solvent-precipitant system.

From the above results it may be concluded that separation by the precipitation technique is effective if it is carried out in such manner so that poly-A and poly-A-B graft copolymers are coprecipitated and only poly-B is left in solution. If a θ-solvent which has widely different θ-temperatures is available for the two homopolymers, the modified precipitation technique is very advantageous because solution and precipitation of polymers can be done merely by raising and lowering the solution temperature.

B. Extraction

In the grafting onto fibers or films, the homopolymer occluded in the grafting product is generally removed by extraction with a selective solvent[6]. This method is in itself relatively simple and hence the most widely used, though time-consuming. However, it is pointed out that an appreciable amount of homopolymer is furthermore extracted if the unreacted substrate polymer has been previously removed. According to Stannett and collaborators[7], the conventional extraction method is inadequate to obtain the pure graft copolymer free of homopolymers, unless both homopolymers are subjected to a repeated, alternate extraction. We have also revealed that the alternate extraction is necessary for separation in various cases[8]. However, there are serious problems with respect to the extractability of homopolymers.

One is the possibility that the graft copolymer is also "dissolved" away as micelle into the extracting solvent. If this actually occurs, the extraction method cannot be applied.

1. Dispersibility of Graft Copolymers into Selective Solvents

Because in the same molecule a graft copolymer has two sequences different in their solubility behavior, it seems possible that the graft copolymer is dispersed into the extracting solvent, resulting in micelle formation[9]. In fact Kotaka *et al.*[10] and Uchida *et al.*[11] observed that A-B type block copolymers could pass spontaneously from a dried state into a dispersion when immersed in their selective solvents.

We have studied the dispersibility of several pure PVAc-styrene graft copolymers with one PS branch in various selective solvents mainly at room temperature[5]. The experiment was done with two kinds of dried samples; one was recovered from a tetrahydrofuran solution by pouring it into water and the other from a benzene solution which was poured into n-hexane. Let us refer to the former sample as A and the latter sample as B. Due to the difference in solubility of each polymer sequence in those solvents, sample A is supposed to have approximately such a microstructure that PVAc chains are extended and PS chains collapsed, while sample B has the inverse structure. A similar tendency was also pointed out by Merrett[12]. The results are summarized in Table 2.

Table 2. Dispersibility of PVAc-styrene graft copolymers in various solvents from dried samples at room temperature; graft copolymer/solvent = 1/100 (w/v)

Solvent	PVAc[a]	PS[b]	Graft copolymer	
			A	B
Methanol	soluble	insoluble	ND	ND
CH[c]	insoluble	T_θ = 34 °C	ND	ND
Acetone	soluble	swollen	D	D
EAA[d]	soluble	T_θ = 108.5 °C	D	D
OAc[e]	T_p = 83 °C[f]	soluble	ND	D

A: Graft copolymer recovered by pouring the dilute THF solution into water.
B: Graft copolymer recovered by pouring the dilute benzene solution into n-hexane.
D: Dispersed, ND: Not dispersed.
[a] $\bar{M}_n = 1.14 \times 10^5$.
[b] $\bar{M}_n = 2.11 \times 10^5$.
[c] Cyclohexane (at 50 °C).
[d] Ethyl acetoacetate.
[e] n-Octyl acetate.
[f] Initial cloud point.

It is obvious from the table that neither sample A nor B could be dispersed in the solvents in which one chain is soluble but the other is completely insoluble (methanol and cyclohexane). On the other hand, in a solvent which can swell the insoluble chain to a significant extent (acetone), both samples A and B are completely dispersed and the solution appears opalescent to reflected light. A similar appearance is observed in ethyl acetoacetate and n-octyl acetate which are θ-solvents for less soluble chains and good solvents for others. However, in n-octyl acetate sample A is hardly dispersed, though sample B is completely dispersed. Since n-octyl acetate is a selective solvent, similar to ethyl acetoacetate except that the PS branch is soluble in the former while the PVAc backbone is soluble in the latter, it cannot be ignored that the sequential structure of the graft copolymer influences its dispersibility in these θ-solvents. Anyway, if one of the component polymers is quite insoluble in the selective solvent, it is clear that the dispersion does not occur, regardless of the sequential structure of the graft copolymer molecule and the microstructure of its bulk sample.

As a conclusion one can say that the graft copolymer with at least one branch can be isolated by the extraction method with the selective solvent, which is a good solvent for the one chain but a very bad solvent for the others. Even if dispersion into the extracting solvent occurred as is frequently observed in the isolation of the graft copolymers with numerous branches[5)] or A-B type block copolymers[10)], they may be freed of the homopolymers by decreasing the solvency of the extracting solvent for the soluble chain, for example, by adding a precipitant for both homopolymers until the copolymer micelle has coagulated.

2. Factors Affecting the Extractability

In contrast with mechanically-blended homopolymers, grafting products need to be extracted alternately with each selective solvent and over a very long period of

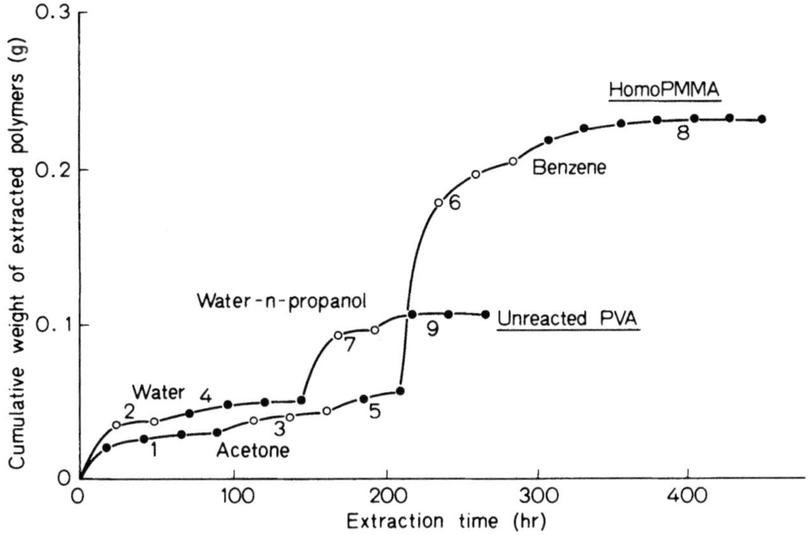

Fig. 9. Alternate extraction of PMMA homopolymer and unreacted PVA from the reaction product, obtained by the mutual irradiation grafting of MMA onto water-swollen PVA films

time. Figure 9 shows a typical example of the alternate extraction for the product obtained by radiation grafting of methyl methacrylate (MMA) onto poly(vinyl alcohol)(PVA) films. The numbers in the figure denote the order of the extraction. As is obviously seen, it is essential for the effective removal of the homopolymers that the alternate extraction be repeated several times. Since the molar mass of the PMMA homopolymer is as high as several million in this case, better extracting solvents than acetone and water, such as a benzene and water-n-propanol (75 : 25) mixture, are required for effective extraction. For the sample consisting of PMMA with the average molar mass ($\bar{M}_n = 1.8 \times 10^5$) comparable to that of PVA, the extracting is more complicated; the extract solution for PVA becomes milky and the separation into two phases does not occur even after centrifugation. This means that the PVA-MMA graft copolymer whose branch has the length comparable to that of the backbone can be dispersed as micelle into water-n-propanol. Therefore, in this case water should be used instead of a water-n-propanol mixture as the solvent to extract the unreacted PVA.

On the other hand, similar alternate extraction gives an erroneous result for the reaction product obtained by the grafting of styrene onto PVA (extractions 1–7 in Fig. 10), unless the PVA part in the product is acetylated to PVAc. The acetylation is carried out for the residue taken out in the middle stage of the 2nd extraction (at A in Fig. 10) and the PVAc and PS are extracted with methanol and cyclohexane, respectively. The result is also given in Fig. 10 (broken line). As can be seen, the acetylation leads to a successful removal of the homopolymers. The removal of the homopolymers from the unacetylated reaction product appears to be accomplished at the 7th extraction, but extraction 3′ for the acetylated sample indicates that the unreacted PVA still remains unextracted in the 7th extraction residue by an amount nearly equal to that of the PVA component in the graft copolymer.

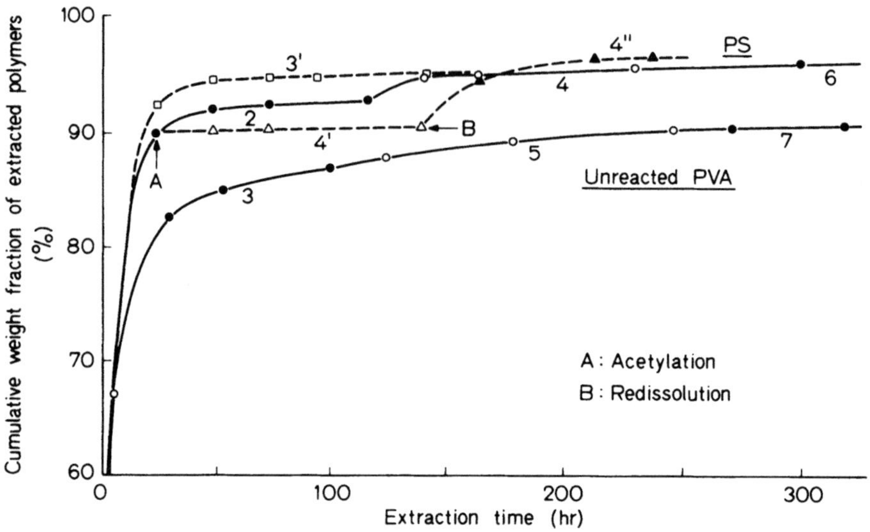

Fig. 10. Alternate extraction of PS homopolymer and unreacted PVA (or acetylated) from the reaction product obtained by the mutual irradiation grafting of styrene onto water-swollen PVA films (The weight of extracted PVAc was plotted in this figure as that of PVA)

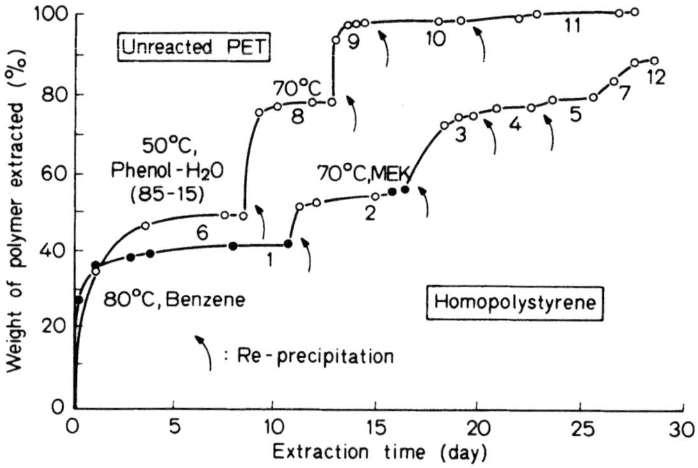

Fig. 11. Extraction of PS homopolymer and unreacted PET from the reaction product obtained by the preirradiation grafting of styrene onto PET fibers

A similarly poor efficiency in extraction can also be seen in Fig. 11, where the extraction result for the product obtained by the radiation grafting of styrene onto poly(ethylene terephthalate)(PET) fibers[13] is shown. In this case the unreacted PET can be extracted after most of PS homopolymer has been extracted by repeated solution-precipitation of the sample. It should be mentioned here that the solution procedure is also necessary to extract the unreacted PET. Such a solution-precipitation procedure is also necessary for nylon-styrene[13] and poly(vinyl chloride)-acrylonitrile[14] graft products so as to remove the homopolymers to a sufficient extent.

Difficulties in the extraction may be chiefly attributed to the microstructure of the sample to be extracted, because the efficiency of extraction may depend on whether the polymer molecule to be extracted is extented or collapsed in the outer component phase[15]. In the experiment shown in Fig. 10, an extraction of PS from the acetylated sample is attempted with cyclohexane at 50 °C, after the PVAc homopolymer is removed. However, the PS homopolymer can hardly be extracted, as shown by curve 4′. Since this sample is recovered by pouring the acetylation mixture (*i.e.*, the pyridine-acetic anhydride solution) into cold water, it may have a microstructure as the PS chain is collapsed, being occluded in the continuous phase of the extended PVAc. Therefore, to inyert the microstructure of the sample it should be dissolved in tetrahydrofuran again and recovered by pouring the solution into n-hexane, whose non-solvency is much stronger for PVAc than PS (at B in Fig. 10). Extraction is then carried out with cyclohexane at 50 °C. The result is shown by curve 4″ in Fig. 10. It is clear that the extended PS homopolymer can be removed further by this extraction, as expected.

Thus for the successful removal of homopolymers it is necessary to take into consideration the microstructure. Alteration of the microstructure by the solution-precipitation may additionally destroy the crystallite or the intimate entanglement among polymer molecules, which makes the extraction difficult.

IV. Contamination of Homopolymers in Isolated Graft Copolymers

There are very few methods of estimating the purity of the isolated graft copolymer. One of these is density-gradient ultra-centrifugation, which requires, however, a complicated technique, and the selection of solvent pairs to produce a necessary density gradient[16]. Clearly, the turbid metric titration method is not appropriate, if the amount of contaminating homopolymer is relatively small. UV, IR and NMR spectroscopy, though utilized by many workers, are by no means effective in determining the purity.

Recently it has been reported that thin-layer chromatography (TLC) makes it possible to characterize polymers with respect to the differences in composition[17–22]. Thus TLC seems to be a suitable technique for estimation of the purity of graft copolymers. The adsorption and desorption of polymers on the chromatographic gel is strongly dependent on the polarity of the solvent and the polymer as well as the activity of stationary gel layer. In general, a mixture of polymers with largely different polarities is readily separated in TLC, since the less polar polymer is developed while the more polar one remains at the spotted point on the plate. In fact, it is impossible to develop the more polar polymer alone, as long as the separation proceeds by an adsorption-desorption mechanism. In order to develop only the more polar polymer, chromatography which proceeds through a selective elution mechanism is required. The developer to be used for such chromatography must be a solvent for the more polar polymer but a nonsolvent for the less polar one and, of course, must be capable of developing the former. Furthermore, the continuous development method is often required to get a satisfactory separation.

We describe below two examples of TLC, applied to the graft copolymers isolated from products of radiation-graft polymerization[23]. The graft copolymers used are PVA-styrene[24] and cellulose-styrene[25]. Prior to the chromatographic development, the homopolymers are removed as rigorously as possible by extraction or selective precipitation. The hydroxyl groups in the graft copolymers are completely acetylated to obtain PVAc-styrene graft copolymer and cellulose triacetate (CTA)-styrene graft copolymer. The characterization study proved that all the graft copolymers have one branch per molecule on the average, the number-average molar mass of the branches being 1×10^5. The homopolymers used are the mother polymers and those formed during the grafting. The TLC substrate is silica gel precoated on glass plates. Between 4 to 40 μg of the polymers are deposited by using a microsyringe as a band of 5 cm length parallel to the developing direction. This band-like deposition is made because about 40 μg are necessary to detect amounts as small as <1% homopolymers occluded in the sample, and such a large amount of polymer normally cannot be developed when deposited as a tiny spot. The developers are listed in Table 3, together with solvents used for preparing sample stock solutions and indicators for the staining of chromatograms.

Typical chromatograms of PVAc-styrene graft copolymer are demonstrated in Figs. 12 and 13. The TLC is always carried out not only for the graft copolymer sample, but also for mixtures of known quantities of sample and homopolymer in order to evaluate quantitatively the amount of the homopolymer contaminating the starting graft copolymer sample. Figure 14 shows the scanning spectrodensitometric traces with reflected light at 256 nm for PS in the developments of mixtures of PS and PVAc-styrene graft copolymer. It seems highly possible that the graft copolymer is not at all developed, the corresponding homopolymer alone being developed.

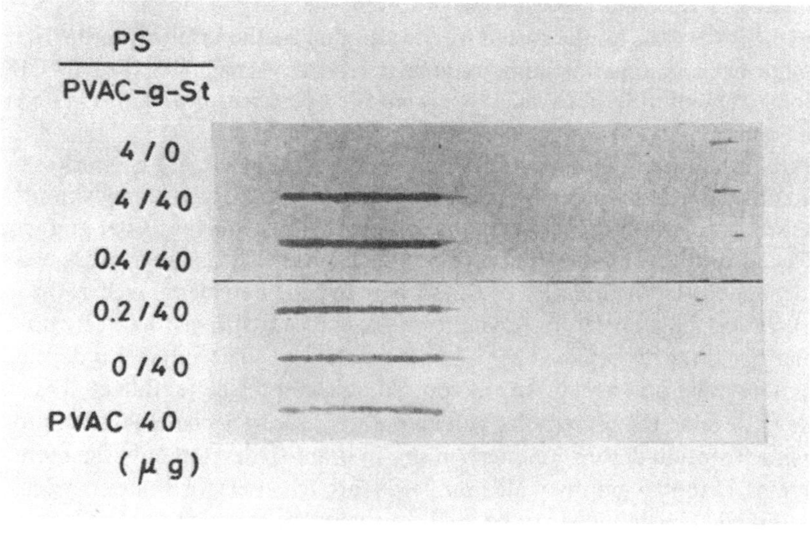

Fig. 12. Thin-layer chromatograms of the PS homopolymer, the PVAc homopolymer, the isolated PVAc-styrene graft copolymer (PVAc-g-PS) and mixtures of PS and PVAc-g-PS with different mixing ratios. Developer: chloroform

Table 3. Experimental conditions of thin-layer chromatography for graft copolymers

Sample	Spotting solvent	Homopolymer corresponding to the backbone		Homopolymer corresponding to the branch	
		Developer	Indicator	Developer	Indicator
CTA-styrene	$CHCl_3$	$CH_2Cl_2/CH_2OH(1:1)$	10% $HClO_4$ aq.soln	$CHCl_3$	10% $HClO_4$ aq.soln
PVAc-styrene	$CHCl_3$	$CH_3OH/H_2O(9:1)$[a]	10% $HClO_4$ aq.soln	$CHCl_3$	10% $HClO_4$ aq.soln

[a] Continuous development.

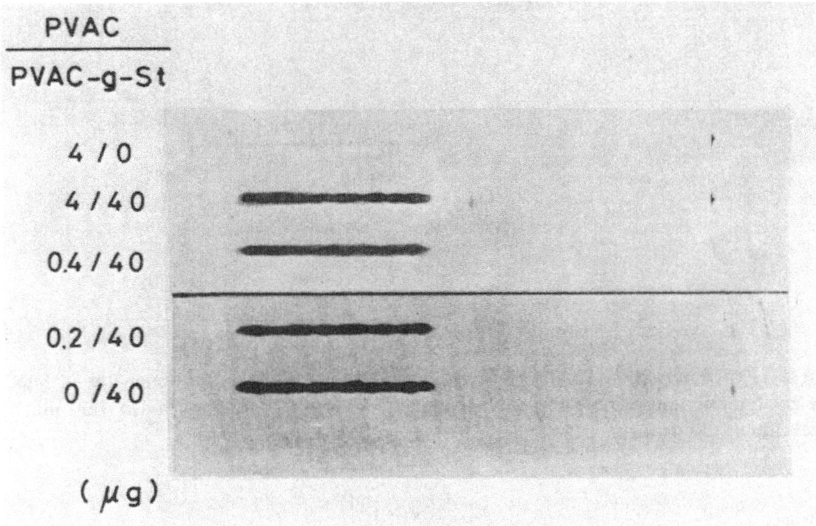

Fig. 13. Thin-layer chromatograms of the PVAc homopolymer, the isolated PVAc-styrene graft copolymer (PVAc-g-PS) and mixtures of PVAc and PVAc-g-PS with different mixing ratios. Developer: methanol-water (9 : 1) mixture

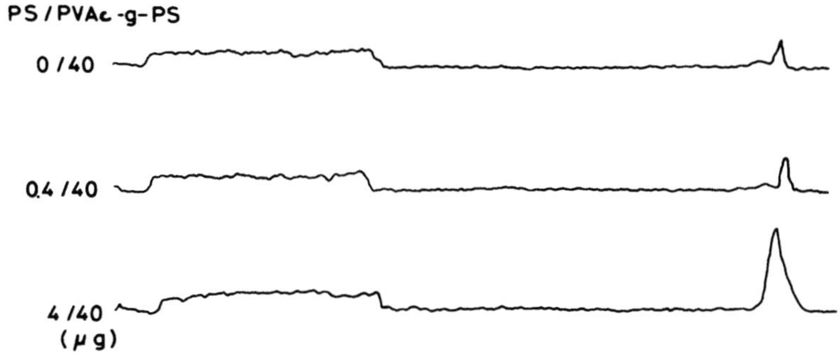

Fig. 14. Scanning spectrodensitometric traces for PS in the chromatograms of the mixtures of PS and PVAc-styrene graft copolymer (PVAc-g-PS) (corresponding to Fig. 12)

Comparison of the staining degrees of the developed polymers reveals the purity of the graft copolymer to be about 95–98%. In Figs. 15 and 16 chromatograms of CTA-styrene graft copolymer are shown which are obtained by using chloroform and a methylene chloride-methanol (1 : 1) mixture as the developer, respectively. The TLC for this sample shows that the purity is over 99%.

It is concluded that we can easily detect homopolymers contaminating graft copolymers using the TLC method, even small amounts of 0.5–1.0%, though in

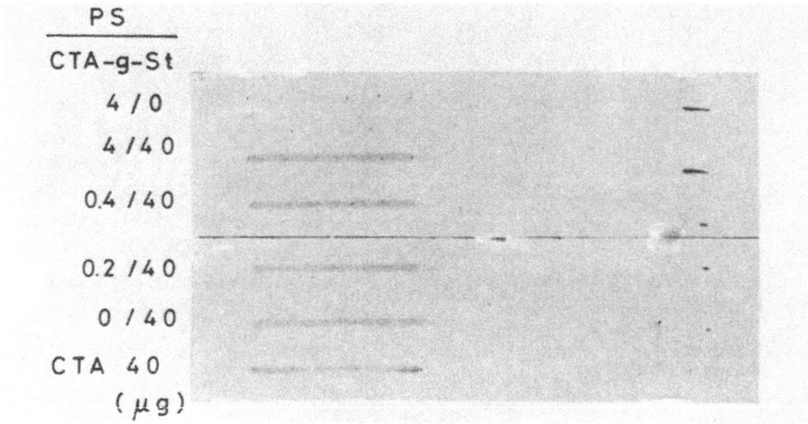

Fig. 15. Thin-layer chromatograms of the PS homopolymer, the CTA homopolymer, the isolated CTA-styrene graft copolymer (CTA-g-PS) and mixtures of PS and CTA-g-PS with different mixing ratios. Developer: chloroform

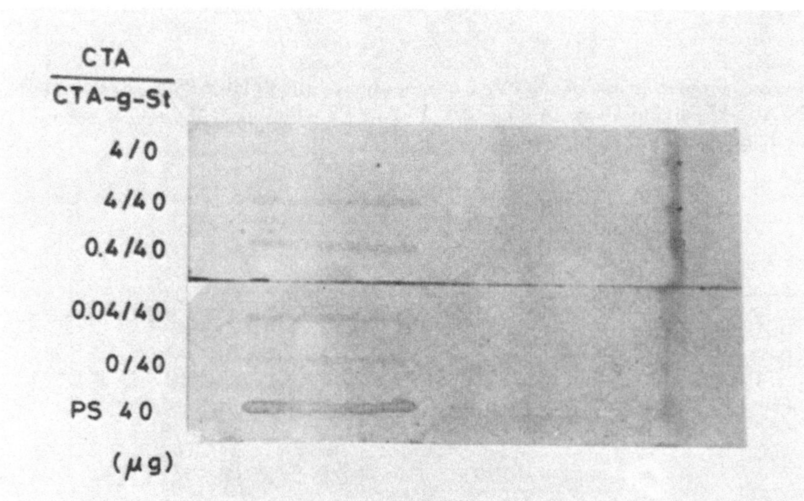

Fig. 16. Thin-layer chromatograms of the CTA homopolymer, the PS homopolymer, the isolated CTA-styrene graft copolymer (CTA-g-PS) and mixtures of CTA and CTA-g-PS with different mixing ratios. Developer: methylene chloride-methanol (1 : 1) mixture

some cases with selective solvents the development should be continued carefully for more than 10 hr. This success in polymer characterization by the TLC method suggests that the grafting product would also be separated on a preparative scale with the help of a similar chromatographic technique. In fact, the pure graft copolymer can be effectively isolated from its reaction product by column adsorption chromatography as described below[26].

Figure 17 shows the chromatogram where the polymer concentration c is plotted against the elution volume. The sample used is a mixture of PVAc-styrene graft copolymer, PVAc and PS. Silica gel is charged with benzene to a column

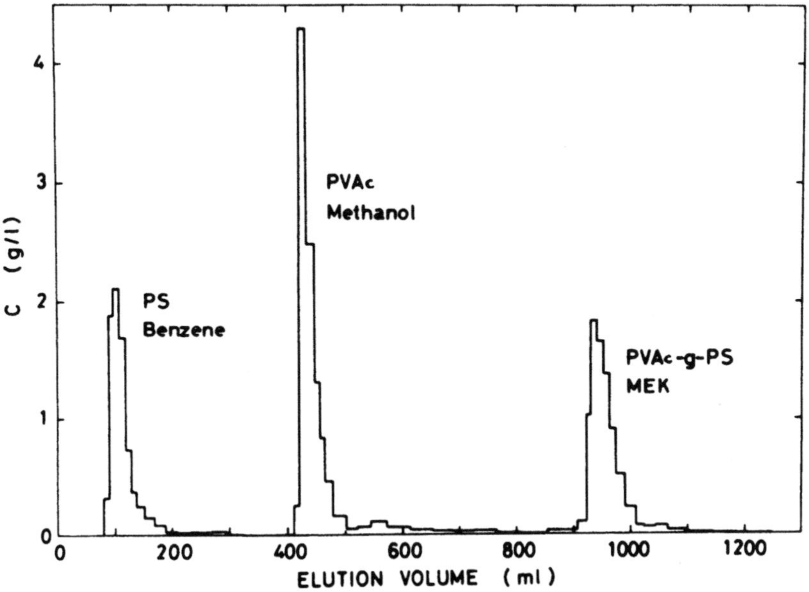

Fig. 17. Chromatogram of grafting product eluted successively using benzene, methanol and MEK as eluent

equipped with a glass jacket. After washing the column with benzene, the polymer solution (308.0 mg/30 ml benzene) is allowed to flow. The solvents used as an eluent are benzene for PS, methanol for PVAc and methyl ethyl ketone (MEK) for the graft copolymer. Evidently, the separations with benzene and MEK are based on the difference in adsorption strength of the polymers to silica gel, whereas that with methanol is based on the difference in the solubility of the polymers for the solvent. In order to avoid mixing of benzene and methanol, CCl_4, with which any of three polymers are not at all eluted, is allowed to flow between the elutions with benzene and methanol. Each solvent is made to flow till the amount of eluted polymer becomes negligible and then replaced by the subsequent solvent.

It seems that the product is expectedly divided into three fractions by this chromatography. Thin-layer chromatographic analysis for the fractions of the graft copolymer indicates that each fraction is not contaminated by more than 0.5–1 wt% with either homopolymer. This contamination could be decreased if the elutions would be continued further with the respective solvents. It is found from IR spectra that PS fractions are virtually uncontaminated with PVAc, while initial PVAc fractions contain a PS component in a larger than negligible amount. This may be due to the elution of a small amount of the graft copolymer with CCl_4-methanol mixture formed during the replacement of CCl_4 by methanol. The recovery of the polymers is quite quantitative (308.0 mg used, 308.6 mg recovered), but partial hydrolysis takes place when removal of free acid from the silica gel is insufficient.

V. Chemical Structure of Graft Copolymers

When isolation and purification of a graft copolymer is accomplished, we should undertake to determine the chemical structure of the graft copolymer.

Structural factors characterizing the graft copolymer are basically the number of branches and the length (molar mass) of the backbone and of the branch. The reported values are mostly average values, since the precise determination of their distribution is much more complicated or sometimes impossible.

The average number of grafted branches in one graft copolymer N_g and the number-average molar mass of the backbone $\overline{M}_{B,g}$ are readily determined from Eqs. (29) and (30), resp., if the number-average molar mass of the graft copolymer $\overline{M}_g$ and of one branch $\overline{M}_b$ are known together with the weight fraction of the branch part in the graft copolymer w_b:

$$N_g = \frac{\overline{M}_g \times w_b}{\overline{M}_b} \quad \text{and} \tag{29}$$

$$\overline{M}_{B,g} = \overline{M}_g(1 - w_b) \tag{30}$$

It should be emphasized that contamination of any homopolymers in the graft copolymer leads to an erroneous determination of the values of $\overline{M}_g$ and w_b.

In the following we briefly show three representative examples of the determination of the chemical structure of graft copolymers, which were carried out in our laboratory.

A. Redox Graft Copolymerization[27)]

It has been found that the oxidation of PVA by persulfate produces radicals on the PVA main-chain as an intermediate species[28)]. This is a consequence of a redox reaction between >CHOH of PVA and $S_2O_8^{2-}$, which seems to take place through Eqs. (31)–(33):

$$S_2O_8^{2-} \longrightarrow 2\,SO_4^{\bullet-} \qquad (31)$$

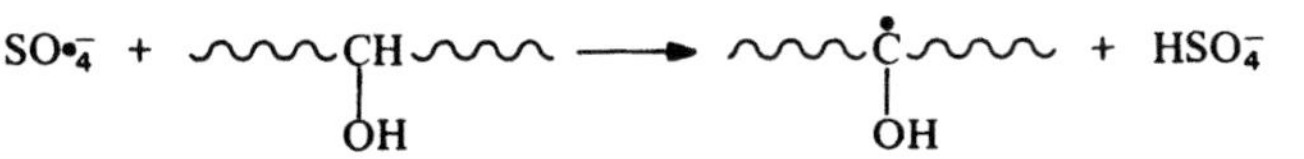

(32)

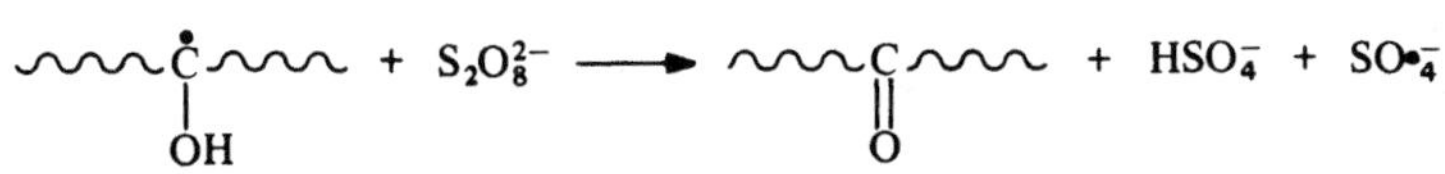

(33)

Therefore graft copolymerization is expected to take place if a vinyl monomer is added to the PVA-persulfate mixture.

We have conducted a graft copolymerization of MMA onto PVA making use of potassium persulfate (KPS) in three different ways: with water-swollen PVA film containing KPS and with PVA solutions containing KPS in H_2O and DMSO. The polymerization of MMA proceeds heterogeneously in the two former cases and homogeneously in the latter case.

The MMA homopolymer and the unreacted PVA are removed from the reaction product by the selective extraction method. The grafted branch PMMA is separated from the backbone by oxidative cleavage of all 1,2-glycols of PVA (about 2 mole%)[29]. Molar masses of the isolated graft copolymers and the separated branches are osmometrically determined, after acetylation of hydroxyl groups in benzene. The chemical composition of the graft copolymers is determined from the saponification value of the acetylated sample.

The conditions and results of graft copolymerization of MMA in aqueous media are given in Tables 4 and 5. Methanol is used to allow the monomer to penetrate into

Table 4. Graft copolymerization of MMA in aqueous media [a]

	Water-swollen film	Aqueous solution
[PVA], wt-%	26.7	3.34
[MMA]/[H_2O] (v/v)	10[b]	0.333
[KPS]/[PVA] x 100 (molar ratio)	0.57	2.0
Polymerization time, hr	5	2
Total conversion of monomer, %	21.2	78.5
Grafting, %[c]	42.8	96.7
Graft efficiency, %[d]	6.12	13.1
Weight fraction of PVA grafted	0.219	0.244
Fraction of KPS participating in the grafting	0.016	0.0054
MMA content of the graft copolymer, mole-%	36.8	63.5

[a] In the absence of air, 60 °C.
[b] Trichloroethylene (TCE) was added; [TCE]/[MMA] = 1.5 (molar ratio).
[c] (Weight of branch PMMA)/(weight of starting PVA) x 100.
[d] (Weight of branch PMMA)/(weight of total PMMA formed) x 100.

Table 5. Characterization of graft copolymers prepared in aqueous media

	Water-swollen film	Aqueous solution
$\bar{M}_n \times 10^{-4}$		
Starting, acetylated PVA	20.0	17.2
Acetylated graft copolymer	37.0	97.5
Separated PMMA branch	14.6	73.2
MMA homopolymer	10.7	74.1
Number of graft branches[a]	1.0	0.9

[a] Number of branches in one graft copolymer molecule.

the PVA film and trichloroethylene to control the chain length of branch PMMA. Clearly, graft copolymerization takes place but the yield is not high, probably because it is a heterogeneous polymerization.

The results of homogeneous grafting in DMSO are summarized in Table 6. It is seen that graft copolymers having about two to four branches are obtained. The weight fraction of the PVA grafted is higher than 0.8 in every case. Side reactions, such as coloration and crosslinking, occur when reaction conditions are severe, as in DG-3 and DG-6. Consequently the length of the backbone of the graft copolymer becomes much larger than that of the starting PVA molecule in these cases. The dependence of the number of branches on the reaction conditions is obscure, but the branch length seems to decrease with increasing temperature and KPS concentration, as is expected from general radical polymerization kinetics.

B. Radiation Graft Copolymerization

The radiation-induced graft copolymerization has attracted attention due to its facility and wide applicability. Although this grafting method has several advantages such as unselective radical formation on any substrate polymers, the polymerization proceeds heterogeneously in the matrix of the substrate, thereby making the total extraction of homopolymers from the graft product very difficult, especially in a mutual radiation graft copolymerization.

As a typical example, a characterization of the product obtained by mutual graft copolymerization of styrene onto cellulose fiber is described in some detail[25)].

High tenacity rayon was employed as cellulose; the grafting method consisted of placing rayon in the bottom of the small reaction tube, adding a monomer mixture of styrene-methanol-water and CCl_4 (five mole% of styrene), degassing by freeze-thaw method, and finally sealing off the tube while under vacuum. The irradiation was carried out at a dose rate of 1.0×10^4 rad/hr. For the isolation of the apparent graft, the reaction product was treated with a large amount of benzene at room temperature to remove the styrene homopolymer formed in the solution, then washed with water and finally extracted with boiling benzene to obtain the apparent graft copolymer. Table 7 shows the experimental results together with that of the

Table 6. Graft copolymerization of MMA onto PVA in DMSO solution by KPS[a]

	Expt. DG-1	Expt. DG-2	Expt.-DG-3	Expt. DG-4	Expt. DG-5	Expt. DG-6
Polym. temperature, °C	60	40	60	60	30	80
[KPS], wt-%	0.31	0.06	0.81	0.06	0.31	0.31
Polymerization time, hr	2	41	2	2	15.5	2
Graft copolymer	Soluble	Soluble	Colored, Insoluble	Soluble	Soluble	Insoluble
MMA content of acetylated graft copolymer, wt-%	58.6	71.7		65.4	71.8	
$\bar{M}_n \times 10^{-4}$						
Acetylated graft copolymer	36.5	109		61.2	85.1	
Separated PMMA branch	5.37	19.9	Lower than $1-2 \times 10^4$	15.2	19.5	
Number of graft branches[c]	3.6	3.9		2.6	3.1	

a [PVA] = 2.5 wt-%; [MMA] = 22.5-%; initial PVA = 2 g; $\bar{M}_n$ of initial PVA = 5.9×10^4.
b The separated PMMA branch permeated a Sartorius membranfilter (UFF Sd) used for the osmometry.
c Number of branches in one graft copolymer molecule.

Table 7. Grafting of styrene onto cellulose by mutual irradiation method. [Cellulose: High tenacity rayon ($\bar{P}_v$ = 293); Monomer mixture: Styr.: MeOH : H_2O = 20 : 72 : 8; polymerization = 50 °C in vacuum, 20 hr (2 x 10^5 rad)]

Transfer agent CCl_4/St.	In mole	0.05	0
Monomer conversion	%	34.3	95
Apparent percent graft[a]	%	44.0	181
$\bar{P}_v$[b] of styrene homopolymer formed in matrix		1,400	12,700

[a] Weight increase after extraction expressed in percent of the weight of the starting cellulose.
[b] Viscosity-average degree of polymerization.

grafting in the absence of carbon tetrachloride. It may be seen that the momomer conversion and the apparent percent graft are remarkably decreased by the presence of CCl_4 · $\bar{P}_v$ of styrene homopolymer formed in the matrix also shows a much lower value when the reaction is carried out in the presence of CCl_4. Therefore, it is expected that the branch and backbone of the graft copolymers obtained in the presence of CCl_4 will have a balanced length.

In order to make the separation of unreacted cellulose from the apparent graft easier it was acetylated under nondegradative conditions in acetic anhydride-pyridine mixtures. The apparent graft copolymer was first soaked in methanol, methanol was replaced by water and then acetylated with a 1 : 2 acetic anhydride-pyridine mixture for 36 hr at 100 °C. After the reaction, the product was precipitated with n-hexane. The precipitate was redissolved in chloroform-methanol and reprecipitated with methanol. The reprecipitation was repeated and finally a purified acetylated product was obtained. When pure cellulose was used for the acetylation it was confirmed that by this procedure cellulose is converted almost totally (98.5 to 99.6%) to triacetate.

It was first intended to remove styrene homopolymer and unreacted cellulose (in the form of triacetate) by alternate extractions with benzene and a 1 : 1 methylene chloride-methanol mixture, but this was not successful. Therefore, a fractional precipitation method was adopted. The acetylated apparent graft copolymer was dissolved in a methylene chloride-methanol mixture (80 : 20 by volume). Methanol was very slowly added to the solution to precipitate the styrene homopolymer and the true graft copolymer. Dissolution in the methylene chloride-methanol mixture and precipitation with methanol were repeated four times. The final solution contained 45.0–46.3% methanol.

The above precipitate was extracted with boiling benzene to remove the homopolymer and finally the true graft copolymer was isolated.

The material balance in the course of the grafting, separation of apparent graft, acetylation and isolation of the true graft is shown in Table 8. Starting with 4.910 g cellulose, 10.666 g acetylated apparent graft copolymer was obtained; when we assume that cellulose is fully converted to triacetate and recovered in the acetylated apparent graft, the calculated weight of the acetylated apparent graft is 10.930 g.

Table 8. Material balance for the isolation of true graft copolymer of cellulose

	Amount in g	
Starting cellulose		4.910
Styrene homopolymer		7.166
formed out of cellulose	(6.427)	
extracted with boiling benzene	(0.739)	
Apparent graft copolymer		7.070
Acetylated apparent graft copolymer		10.666
Styrene homopoly. isolated after acetylation	1.520	10.930 calc.
Extracted cellulose triacetate	7.517	
Isolated true graft copolymer (acetylated)	0.978	
Loss in the processing	0.651	
Sum of the above figures	10.666	

The agreement is satisfactory when we assume a loss of 0.264 g in the separation of the acetylated apparent graft copolymer. The total loss in the course of the isolation of the true graft copolymer was 0.651 g. True percent graft and fraction of cellulose grafted were 11.1 and 0.05 respectively.

The acetylated true graft copolymer was dissolved in methylene chloride, acetone was added to the solution to obtain a 1 : 1 methylene chloride-acetone composition of the solvent, concentrated hydrochloric acid was added to obtain a 3N solution, and hydrolysis was carried out for 72 hr at 60 °C. The hydrolysis proceeded not in a homogeneous but in a highly swollen state. The branch was precipitated by pouring the hydrolysis mixture into methanol. A part of the precipitated branch was dissolved in m-cresol and filtered. The other part was treated with 1 : 2 acetic anhydride-pyridine mixture for 15 hr at 100 °C to acetylate cellulose fragments at the end of polystyrene branches.

Molar mass determination of the acetylated true graft copolymer, the separated branch after acetylation and the acetylated mother cellulose, was carried out with a high speed membrane osmometer using chloroform, benzene, and chloroform, resp., as solvents.

As shown in Table 9, the number-average molar mass of the acetylated true graft copolymer, the separated branch after acetylation and the acetylated mother cellulose, was 306,000, 187,000 and 72,400, respectively. A benzene solution of the styrene homopolymer formed in the matrix of the cellulose exhibited an $\overline{M}_n$ of 195,000, which agrees well with $\overline{M}_n$ of the separated branch. The benzene solution of the separated branch was slightly turbid before acetylation and gave a molar mass of 236,000. Before acetylation, the IR-spectrum of the separated branch exhibited absorptions near the wave number 3400 cm^{-1}, which were probably due to OH-groups. Therefore, a slightly higher molar mass of the separated branch before acetylation may be attributed to partial association.

Table 9. Molar mass and chemical structure of acetylated graft copolymer

a. Molar mass			
Polymer	Solvent, temp.		$\bar{M}_n$
Acetylated true graft copolymer ($\bar{M}_g$)	chloroform	30 °C	306,000
Separated branch after acetylation ($\bar{M}_b$)	benzene	30 °C	187,000
Acetylated mother cellulose ($\bar{M}_{B,O}$)	chloroform	30 °C	72,400
b. Composition			
Composition of acetylated true graft copolymer Styrene : Triacetate = 0.558 : 0.442			
Molar mass of polystyrene part			171,000
Molar mass of triacetate part			135,000
wt fraction of cullulose grafted	0.05		
c. Structure			
Number of branches in one graft copolymer			0.915
$\bar{M}_n$ of the acetylated cellulose backbone of the graft copolymer ($\bar{M}_{B,g}$)			135,000

Chemical composition of the acetylated graft copolymer was determined by saponification, and it was found that the graft copolymer consisted of 0.558 parts of polystyrene and 0.442 parts of triacetate by weight. The number of branches per one graft copolymer and $\bar{M}_{B,g}$ was calculated by Eqs. (29) and (30). The results are shown in Table 9. The number of branches per one graft copolymer is one within the experimental error and $\bar{M}_{B,g}/\bar{M}_{B,O}$ ($= \bar{P}_{B,g}/\bar{P}_{B,O}$) is 1.87. These results are in good agreement with the statistics derived in Chapter II, since the weight fraction of the mother polymer grafted, $\underline{w}$, is only 0.05. As is seen from Eqs. (15) and (21), in such a case, $N_g = 1$ and $\bar{P}_{B,g}/\bar{P}_{B,O} = 2$. Hence it is scarcely possible that crosslinking of cellulose molecules has taken place during the reaction and it is also improbable that termination by the coupling of growing side chains has occurred to a noticeable extent in the presence of a large amount of a chain transfer agent. Also such a low weight fraction of cellulose grafted is consistent with a low radiation yield of cellulose radicals upon irradiation[2].

C. Grafting by Polymer Coupling

In contrast to graft copolymerizations, the coupling grafting utilizes two different prepolymers having functional groups reactive with each other to yield graft copolymers. If both of the prepolymers are monodisperse in molar mass distribution, the characterization of the grafting product may be much easier, except for the isolation of the pure graft copolymer. Even though carried out with the prepolymers of broad molar mass distribution, it seems likely that the lengths of the backbone and branches are close to those of the corresponding prepolymers, if the coupling reaction is allowed to proceed to a high extent. However it should be pointed out that the difference in the chemical composition among the graft copolymer molecules becomes large even in this case[30].

The coupling grafting, which is illustrated below, is effected by condensation between amino groups on PVAc and an acyl chloride endgroup of PS[31]. The PVAc, having a number of pendant amino groups distributed along the mainchain ($PVAc/NH_2$), can be synthesized by the partial acetalization of PVA with amino-acetaldehyde, followed by acetylation of the hydroxyl groups remaining in the polymer. PS, carrying one terminal acyl chloride group (PS-COCl), can be obtained by polymerization of styrene in the presence of trichloroacetyl chloride (TCAC). The condensation coupling reaction between the two polymers is performed at room temperature in a chloroform solution containing triethylamine (TEA). The graft copolymer, isolated by an extensive extraction of the unreacted prepolymers with methanol and cyclohexane, is subjected to determination of the chemical composition and number-average molar mass. Acidic hydrolysis of the acetal at the branching site in the graft copolymer results in the separation of the grafted PS branches from the backbone.

Figure 18 shows the yield of the graft copolymer in the coupling reaction carried out at a total polymer concentration of 5.0 g/dl. Since the $PVAc/NH_2$ prepolymer contains 2.0 mole% of the amino group and the $\bar{M}_n$ of this polymer is 13.0×10^4, 30 amino groups may be contained on the average in one $PVAc/NH_2$ molecule. The $\bar{M}_n$ values of PS-COCl used are 4.62×10^4, 9.22×10^4 and 14.3×10^4.

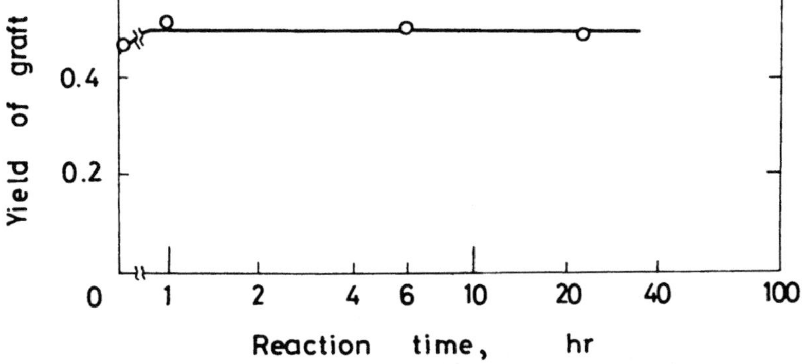

Fig. 18. Influence of reaction time on graft copolymer yield (wt. of $PVAc/NH_2$ = 0.30 g and total polymer conc. = 5.0 g/dl): (○) $\bar{M}_n$ of PS-COCl = 4.62×10^4 and wt. of PS-COCl = 0.60 g; (●) $\bar{M}_n$ of PS-COCl = 9.22×10^4 and wt. of PS-COCl = 1.50 g; (□) $\bar{M}_n$ of PS-COCl = 14.3×10^4 and wt. of PS-COCl = 2.00 g

Thus, the concentration of amino groups in the reaction mixture is high enough to react with all the acyl chloride groups of PS. After the addition of TEA to the reaction mixtures and stirring for periods of 2 min, 1, 6 and 24 hr, hydrogen chloride is added in excess to stop the reaction. It appears that the coupling reaction comes to a completion within about 2 min in all cases. It is noteworthy that the reaction mixture of 5.0 g/dl, before coupling, appears turbid owing to the incompatibility between PS and PVAc, but it becomes clear immediately on addition of TEA.

Figure 19 shows the graft coupling efficiency of PS as a function of the PS/PVAc ratio in the feed and of the reaction time. It is seen that the graft coupling efficiency is in the vicinity of 0.4, irrespective of the polymer mixing ratio and of the $\overline{M}_n$ of PS-COCl.

By applying the theoretical consideration described elsewhere[32] to the polymerization of styrene in the presence of TCAC, we can predict that about 90% of the PS molecules must have terminal acyl chloride groups. Therefore, the finding that the graft coupling efficiency is lower than 0.9 would be the result of the disappearance of acyl chloride groups by hydrolysis due to trace amounts of water in the coupling medium.

The graft coupling efficiency of backbone PVAc/NH_2, plotted against the polymer mixing ratio in Fig. 20, implies that a considerable amount of PVAc carries no PS branches, when the weight ratio of PS-COCl to PVAc/NH_2 in the reaction mixture is as low as unity. This is obvious, because the average number of branches per backbone chain is very low for such mixing ratios, as expected from the statistical calculation in Chapter II. Figure 21 shows the yield of graft copolymer in coupling reactions performed with a constant amount of PVAc/NH_2 and different amounts of PS-COCl. The total polymer concentration in the mixture is kept at 5.0 g/dl and

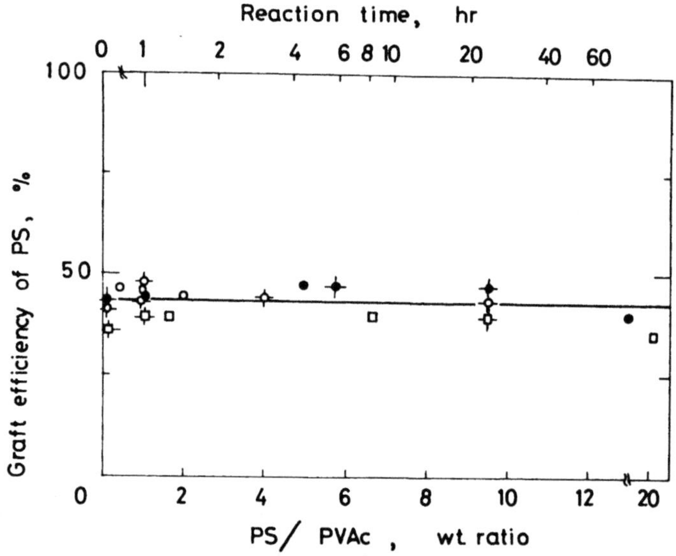

Fig. 19. Influence of the polymer weight ratio of feed and reaction time on the graft efficiency of PS-COCl: plots for PS/PVAc, (○) $\overline{M}_n = 4.62 \times 10^4$, (●) $\overline{M}_n = 9.22 \times 10^4$, (□) $\overline{M}_n = 14.3 \times 10^4$; plots for time, (-○-) $\overline{M}_n = 4.62 \times 10^4$, (-●-) $\overline{M}_n = 9.22 \times 10^4$, (-□-) $\overline{M}_n = 14.3 \times 10^4$

the time of reaction is 24 hr. The yield increases as the added amount of PS-COCl becomes larger.

The characterization results of the product are summarized in Table 10. The number of branches per backbone chain was calculated from Eq. (29) assuming the $\overline{M}_n$ of one branch to be equal to the $\overline{M}_n$ of the starting prepolymer. This assumption is supported by the result described in Table 11, where the intrinsic viscosity $[\eta]$ of

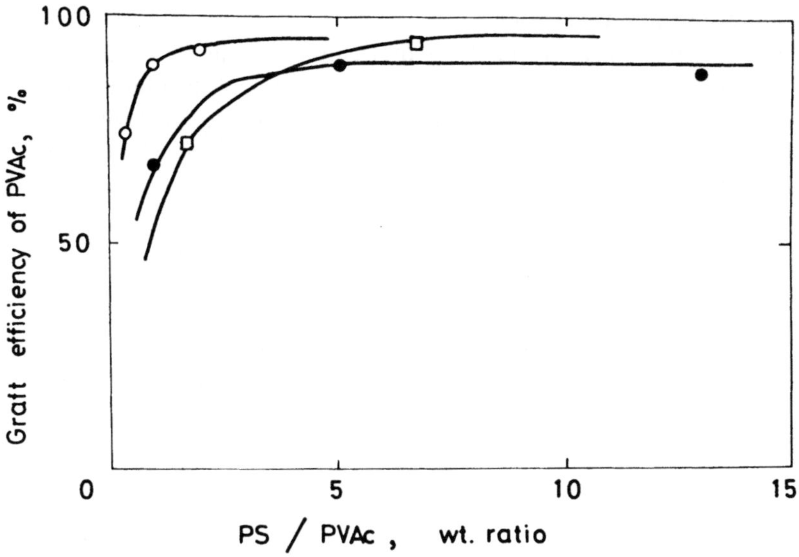

Fig. 20. Influence of the polymer weight ratio of feed on the graft efficiency of PVAc/NH_2: (○) $\overline{M}_n$ of PS-COCl = 4.62 x 10^4; (●) $\overline{M}_n$ of PS-COCl = 9.22 x 10^4; (□) $\overline{M}_n$ of PS-COCl = 14.3 x 10^4

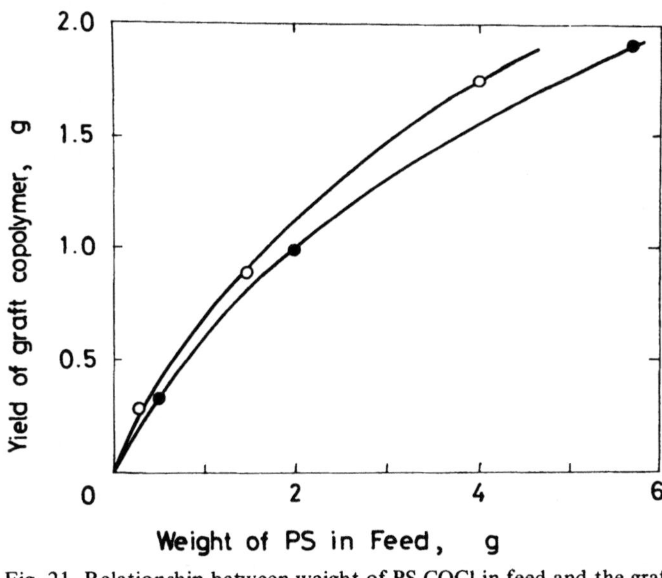

Fig. 21. Relationship between weight of PS-COCl in feed and the graft copolymer yield: (○) $\overline{M}_n$ = 9.22 x 10^4; (●) $\overline{M}_n$ = 14.3 x 10^4

the starting PS is compared with that of a PS branch separated from the graft copolymer. If the high molar mass fraction of PS-COCl remained uncoupled, [η] of the PS branch would become lower than that of the starting PS, but the observed [η] has a reverse trend, as is seen in Table 11. Accordingly the difference in [η] may be ascribed to the loss of the low molar mass fraction of the separated branch polymer during the recovery process. It is noticed in Table 10 that $\bar{M}_n$ of the backbone PVAc in the graft copolymer is higher than that of the starting PVAc/NH_2, being consistent with the statistical calculation.

The number of branches per starting PVAc molecule versus the PS/PVAc ratio in the reaction mixture is plotted in Fig. 22. The good linearity of the plot leads to the conclusion that the PS prepolymers which carry non-deactivated acyl chloride groups all undergo the coupling reaction with PVAc/NH_2, irrespective of the molar mass of the PS molecule, at least in the molar mass range studied.

In Table 12 the number of branches of graft copolymers prepared by coupling reaction using relatively large amounts of PS-COCl and keeping COCl/NH_2 molar ratios in the reaction mixture always to less than unity is summarized. It is seen that a graft copolymer having 15 branches is formed. Alkali hydrolysis of acetyl groups in the graft copolymer produces a graft copolymer consisting of one PVA backbone and many branches of PS. This PVA-styrene graft copolymer is soluble in phenol, a common solvent of PVA and PS.

Table 10. Characterization of the graft copolymers

$\bar{M}_n \times 10^{-4}$ of starting PS	PS/PVAc, wt. ratio	wt-% PS in graft copolymer	$\bar{M}_n \times 10^{-4}$ Graft copolymer	$\bar{M}_n \times 10^{-4}$ Branch part	Number of branches[a]	$\bar{M}_{B,g}$[b] / $\bar{M}_{B,O}$[c]
9.22	1.00	40.6	36.1	14.7	1.60	1.65
9.22	5.00	67.0	55.8	37.8	4.10	1.42
14.3	1.67	46.8	42.7	20.0	1.40	1.75
14.3	6.65	66.9	47.4	31.7	2.11	1.21
14.3	20.0	84.0	103	87.0	5.78	1.25

[a] In one graft copolymer molecule.
[b] $\bar{M}_n$ of backbone part.
[c] $\bar{M}_n$ of starting PVAc (= 13.0 x 10^4).

Table 11. Comparison of length of starting and separated PS

	$\bar{M}_n \times 10^{-4}$	[η][a]	$\bar{M}_n \times 10^{-4}$	[η][a]	$\bar{M}_n \times 10^{-4}$	[η][a]
Starting PS	14.3	0.617	9.22	0.387	4.62	0.260
Grafted PS	15.8	0.718		0.397		0.272

[a] In chloroform, 25 °C.

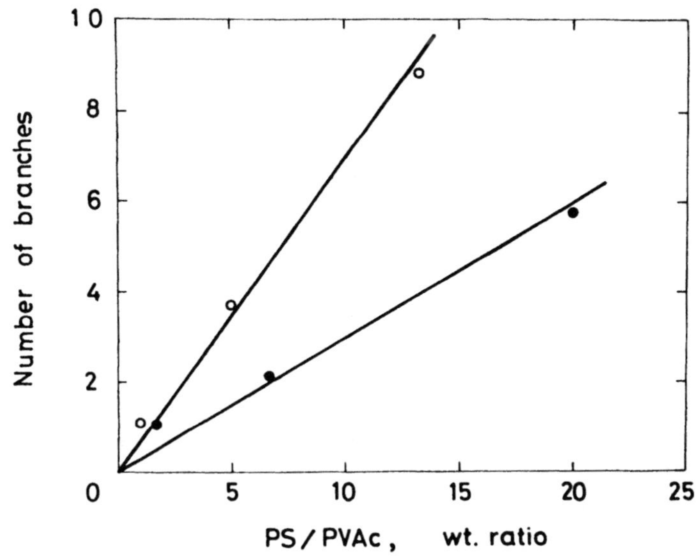

Fig. 22. Number of grafted branches formed per starting PVAc molecule (total polymer conc. = 5.0 g/dl and reaction time = 24 hr): (○) $\bar{M}_n = 9.22 \times 10^4$; (●) $\bar{M}_n = 14.3 \times 10^4$

As demonstrated above, the coupling reaction has proved to be capable of grafting many branches to the backbone polymer merely by varying the ratio of the two prepolymers in the reaction mixture.

VI. Surface Grafting

Less attention has been paid to surface grafting, defined as the grafting taking place only on the surface or in the vicinity of the surface of bulk polymeric substances such as films, fibers and plastics, than to the graftings stated above; but this is none-

Table 12. Other examples of coupling reactions of partially aminoacetalized PVAc with PS having an acyl chloride end group (25 °C, 24 hr, NH_2 content in PVAc = 2.0 mole%)

Code no.	$\bar{M}_n \times 10^{-4}$ of starting PS	PS/PVAc, wt. ratio	wt-% PS in graft copolymer	$\bar{M}_n \times 10^{-4}$		Number of branches[a]
				Graft copolymer	Branch part	
CG-2	0.82	3.30	33.9	33.2	11.3	14
CG-1	0.94	1.50	61.3	22.9	14.1	15
CG-9	2.03	5.00	49.2	53.0	26.1	13
CG-7	8.95	25.0	79.0	53.0	41.9	4.7
CG-10	8.90	15.0	68.3	71.7	48.9	5.6
CG-2	12.3	3.0	63.1	44.3	27.9	2.3

[a] In one graft copolymer molecule.

theless important especially in polymer modifications. Generally the surface grafting involves the polymer reaction through which another macromolecule is chemically connected to a polymeric surface. Since the yield of graft copolymers is extremely small due to surface reactions, it is virtually impossible to characterize the surface-grafting product in the usual fashion. Even in the surface grafting, however, it may be required to provide at least qualitative evidence for the occurrence of the grafting and, if possible, to determine the graft yield. It is also an important problem to clarify whether the grafting is localized essentially on the surface of the substrate polymer or penetrates more or less into its interior.

The methods widely used in characterizing the surface grafting onto films are UV, attenuated total reflection (ATR) IR, X-ray-photoelectric spectroscopy (ESCA), contact angle measurement and other techniques applicable to the characterization of the surface treatments of polymeric substances.

As an example the characterization of surface grafting onto an ethylene-vinyl alcohol (33 : 67) random copolymer (EVA) film[33] is described here. Since this water-insoluble film has hydroxyl groups on the surface, surface grafting may occur, for instance, with the so-called dialdehyde starch (DAS), whose chemical structure is given by

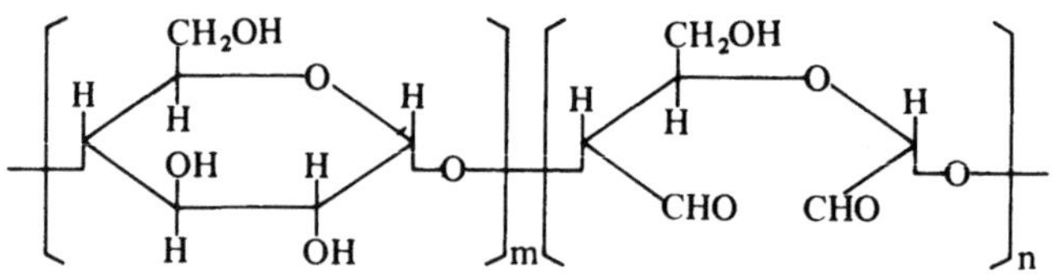

The grafting reaction is expected to take place in acidic aqueous media *via* acetalization between OH groups on the EVA film and the CHO groups of the water-soluble DAS. In Fig. 23 the contact angle of the grafted film against water is shown as a function of time. A decrease in the contact angle, in other words, an increase in

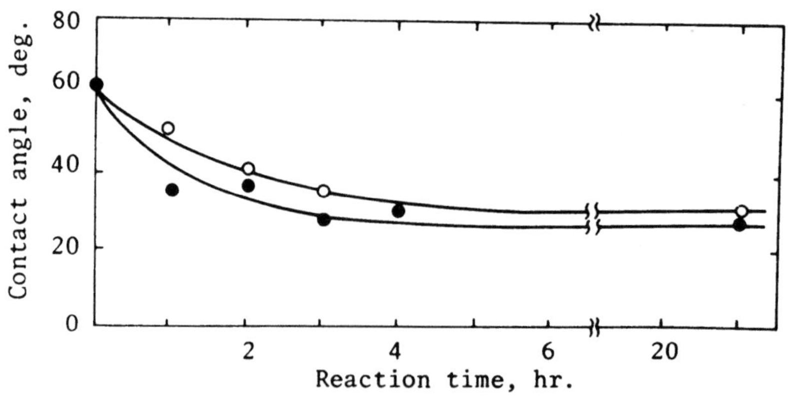

Fig. 23. Change of contact angles of the EVA films by acetalization with DAS ($[DAS]_0$ = 5.0 wt%, [HCl] = 1 N 50 °C): (○) DAS-45; (●) DAS-83 (Degree of oxidation is 45 and 83 mole%)

hydrophilicity of the film surface indicates that grafting is actually occurring. The reaction seems to be restricted to the surface of the film when the following facts are considered;

(1) no detectable weight increase,
(2) no detectable difference in ATR IR spectra before and after the reaction,
(3) low swelling of the film in water (the equilibrated water fraction of the film is 0.06, regardless of the reaction) and
(4) extremely low diffusability of the DAS molecules into the film owing to its high molar mass ($\bar{M}_n$ = 11,000).

The DAS-grafted EVA film can further react with polymers having functional groups capable of coupling to the CHO groups, which are still present in an unreacted state in the DAS molecule. Among these functional groups, NH_2 is highly reactive to CHO forming a Schiff's base. Indeed, proteins, chitosan and polyacrylamide have been successfully grafted onto the DAS-grafted EVA films. The ESCA spectrum of the film, grafted with glycol chitosan, is given in Fig. 24. The starting DAS-grafted film which does not contain nitrogen exhibits, of course, no peak in the range of the binding energy specific to N_{1s}, while a peak is clearly seen at the binding energy corresponding to N_{1s} in the ESCA spectrum for the film further grafted with glycol chitosan.

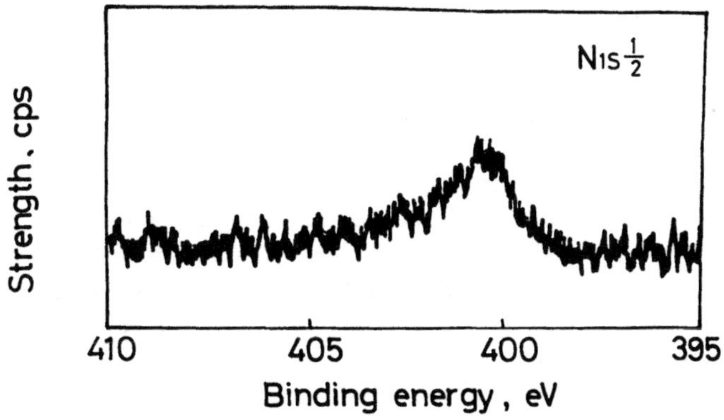

Fig. 24. X-ray photoelectron spectrum of EVA-DAS film grafted with glycol chitosan

Figure 25 shows the result of the surface grafting of four typical proteins onto the DAS-grafted EVA film. The amount of the proteins grafted can be readily determined by the conventional ninhydrin method with high accuracy, but it is difficult to estimate to what extent the native proteins have suffered from denaturation in the course of the grafting. When the CHO groups of the film have been reduced to CH_2OH prior to the grafting, any indication of grafting is unrecognizable. The maximum grafting yield and the dimension of one native protein molecule imply that the graft surface consists of a monomolecular protein layer. The film grafted with amylase is able to hydrolyze amylose in an aqueous solution, again giving evidence for grafting. However, the enzymatic activity is much lower compared with that of the free amylase of the same amount[34].

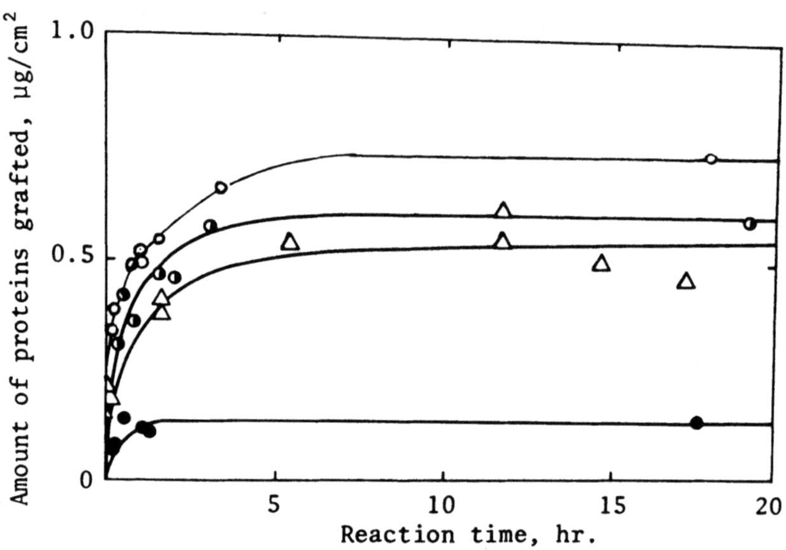

Fig. 25. Grafting of various proteins onto EVA-DAS film surfaces (30 °C): (○) Bovine serum albumin, 0.1 mg/ml, (◑) Fibrinogen, 6.7 mg/ml, 0.26 M NaCl, (△) Amylose, 0.2 mg/ml, 0.01 M $Ca(Ac)_2$, (●) Gelatin, 5 mg/ml, 0.05 M NaOH

VII. Conclusion

This review article has not dealt with all the subjects relating to the characterization of graft copolymers, but only with some of the fundamental ones which must first be taken into account when studying graft copolymers. There is no universal method applicable to the purification and characterization of any graft copolymers. This is not due to the infancy of this field but to the fact that they stongly depend on the synthesis method and the nature of the polymers composing the backbone and the graft chains. Peculiar properties such as emulsifying ability of graft copolymers should also be taken into consideration in their characterization.

Graft copolymerizations have been often employed in industries of plastics, rubber, fiber, adhesives and so on. In most of these cases the gross polymerization product is subjected to processing without isolating the pure graft copolymer. It is well known that the physical properties of the product are closely related to its superstructure, which in turn is a function of both the content and the chemical structure of the graft copolymer present in the product. Therefore, the characterization of the graft copolymer is also desirable to control the product property.

Strictly speaking, studies on graft copolymerization should always be accompanied by the characterization of the graft copolymer formed; at least a determination of the yield of the graft copolymer. If the characterization is not performed at all, one may claim that the grafting product might be merely a mixture of two homopolymers. We have demonstrated that the weight increase of PVA films by a conventional graft copolymerization of MMA is mostly attributed to the MMA homopolymer occluded in the film without being extracted[8]. It is evident that spectroscopy such as UV, IR and NMR are not able to provide any evidence for grafting, unless they can detect the chemical bond newly formed as a result of grafting. Except for the graft copolymer with many branches, the detection is highly difficult because the number of these bonds is too small.

To synthesize graft copolymers for a study of their physico-chemical properties, the polymer coupling method is preferred to the graft copolymerization, though the latter does not need any special techniques. This may be simply because the graft copolymer synthesized by a conventional graft copolymerization is much more difficult to characterize and graft copolymers with well-defined structure can be much more easily prepared by the polymer coupling reaction.

Finally it is interesting to note that naturally-occurring graft copolymers, glycoprotein and proteoglycan, have recently attracted much attention because of their importance in biological science. However, very few glycoproteins have been successfully purified and characterized, but we believe that their characterization studies may make great progress in the near future[35]. Also the graft copolymers which are prepared by the coupling of biologically active substances such as proteins, hormones and polysaccharides with natural or synthetic substrate polymers seem to be increasing in importance[36]. One of the examples is immobilization of enzymes. In these cases, determination of the chemical bond connecting the two different polymers as well as of the configuration and denaturation of the biopolymer grafted should be included in the characterization.

Acknowledgement. The author is indebted to Prof. I. Sakurada for his advice and encouragement and Dr. F. Horii for his discussions and collaboration throughout the course of this investigation. Prof. S. Okamura's invitation to prepare the review and his helpful suggestions are much appreciated.

VIII. References

1. Ikada, Y., Horii, F.: Makromol. Chem. *175*, 227 (1974)
2. Ikada, Y., Horii, F., Nishizaki, Y., Kawahara, T., Uehara, H.: Macromolecules *8*, 276 (1975)
3. Magat, E. E., Millar, I. K., Tanner, D., Zimmerman, J.: J. Polym. Sci., C, *4*, 615 (1963)
4. Ikada, Y., Horii, F., Sakurada, I.: J. Polym. Sci. Polym. Chem. Ed. *11*, 27 (1973)
5. Horii, F.: Ph. D. Dissertation, Kyoto University 1974
6. Battaerd, H. A. J., Tregear, G. W.: Graft copolymers. New York: Interscience 1967
7. Yasuda, H., Wrag, J. A., Stannett, V.: J. Polym. Sci., C, *2*, 387 (1963)
8. Sakurada, I., Ikada, Y., Horii, F.: Makromol. Chem. *139*, 171 (1970)
9. Horii, F., Ikada, Y., Sakurada, I.: J. Polym. Sci. Polym. Chem. Ed. *12*, 323 (1974)
10. Kotaka, T., Tanaka, T., Inagaki, H.: Polym. J. *3*, 327 (1972)
11. Uchida, T., Soen, T., Inoue, T., Kawai, H.: J. Polym. Sci. A-2 *10*, 101 (1972)
12. Merrett, F. M.: J. Polym. Sci. *24*, 467 (1957)
13. Sakurada, I., Ikada, Y., Kawahara, T.: J. Polym. Sci. Polym. Chem. Ed. *11*, 2329 (1973)
14. Kwang-Fu, Chon, Ikada, Y., Sakurada, I.: unpublished work
15. Ceresa, R. J.: Block and graft copolymers. London: Butterworths 1962

16. Ende, H. A., Stannett, V.: J. Polym. Sci., A, *2*, 4047 (1964)
17. Inagaki, H., Matsuda, H., Kamiyama, F.: Macromolecules *1*, 520 (1968)
18. Belenkii, B. G., Gankina, E. S.: Dokl. Akad. Nauk SSSR *186*, 857 (1969)
19. Belenkii, B. G., Gankina, E. S.: J. Chromatogr. *53*, 3 (1970)
20. Kamide, K., Manabe, S., Osafune, E.: Makromol. Chem. *168*, 173 (1973)
21. White, J. L., Salladay, D. G., Quinsenberry, D. O., MacLean, D. L.: J. Appl. Polym. Sci. *16*, 2811 (1972)
22. Kotaka, T., White, J. L.: Macromolecules *7*, 106 (1974)
23. Horii, F., Ikada, Y., Sakurada, I.: J. Polym. Sci. Polym. Chem. Ed. *13*, 755 (1975)
24. Ikada, Y., Horii, F., Sakurada, I.: Bull. Inst. Chem. Res., Kyoto Univ. *49*, 6 (1972)
25. Sakurada, I., Ikada, Y., Nishizaki, Y.: J. Polym. Sci., C, *37*, 265 (1972)
26. Horii, F., Ikada, Y.: J. Polym. Sci. Polym. Letters Ed. *12*, 27 (1974)
27. Ikada, Y., Nishizaki, Y., Sakurada, I.: J. Polym. Sci. Polym. Chem. Ed. *12*, 1829 (1974)
28. Ide, F., Nakatsuka, K., Tamura, H.: Kobunshi Kagaku *23*, 45 (1966)
29. Sakurada, I., Ikada, Y., Uehara, H., Nishizaki, Y., Horii, F.: Makromol. Chem. *139*, 183 (1970)
30. Kotaka, T., Donkai, N., Min, T. I.: Bull. Inst. Chem. Res., Kyoto Univ. *52*, 332 (1974)
31. Ikada, Y., Maejima, K., Iwata, H.: Makromol. Chem. *179*, 865 (1978)
32. Ikada, Y., Iwata, H., Nagaoka, S.: Macromolecules *10*, 1364 (1977)
33. Ikada, Y., Iwata, H., Nagaoka, S.: 26th IUPAC Symposium, Abstracts, P.479, Tokyo 1977
34. Ikada, Y., Mita, T., Iwata, H.: unpublished work
35. Kornfeld, R., Kornfeld, S.: Ann. Rev. Biochem. *46*, 217 (1976)
36. Hixson, H. F., Jr., Goldberg, E., (eds.): Polymer grafts in biochemistry. J. Macromol. Sci. Chem. *A10*, Nos. 1 & 2 (1976)

Received February 21, 1978
S. Okamura (editor)

Preparation and Study of Block Copolymers with Ordered Structures

Bernard R. M. Gallot

Centre de Biophysique Moleculaire, C.N.R.S., 1A, Av. de la Recherche Scientifique, 45045-Orleans, France

Table of Contents

List of Symbols

SB	Polystyrene-polybutadiene
BMS	Polybutadiene-poly(α-methyl styrene)
BVN	Polybutadiene-poly(vinyl naphthalene)
SBS	Polystyrene-polybutadiene-polystyrene
BSB	Polybutadiene-polystyrene-polybutadiene
SI	Polystyrene-polyisoprene
SIS	Polystyrene-polyisoprene-polystyrene
IV2P	Polyisoprene-poly(vinyl-2-pyridine)
IV4P	Polyisoprene-poly(vinyl-4-pyridine)
IMMA	Polyisoprene-poly(methyl methacrylate)
Kraton	Polystyrene-poly(butadiene or isoprene)-polystyrene
Europrène	Star polystyrene-polybutadiene with 4 branches
$(BS)_n$	Star polybutadiene-polystyrene with n branches
$(SB)_n$	Star polystyrene-polybutadiene with n branches
$(SI)_n$	Star polystyrene-polyisoprene with n branches
SIV2P	Polystyrene-polyisoprene-poly(vinyl-2-pyridine)
SV2P	Polystyrene-poly(vinyl-2-pyridine)
SV4P	Polystyrene-poly(vinyl-4-pyridine)
V2P V4P	Poly(vinyl-2-pyridine)-poly(vinyl-4-pyridine)
MMA HMA	Poly(methyl methacrylate)-poly(hexyl methacrylate)
MMA LMA	Poly(methyl methacrylate)-poly(lauryl methacrylate)
MMA OMA	Poly(methyl methacrylate)-poly(octadecyl methacrylate)
HMA LMA	Poly(hexyl methacrylate)-poly(lauryl methacrylate)
SEO	Polystyrene-poly(ethylene oxyde)
BEO	Polybutadiene-poly(ethylene oxyde)
EMAEO	Poly(ethyl methacrylate)-poly(ethylene oxyde)
SCL	Polystyrene-poly(-coprolactone)
BG	Polybutadiene-poly(benzyl-L-glutamate)
SG	Polystyrene-poly(benzyl-L-glutamate)
SC	Polystyrene-poly(cinnamyl-L-glutamate)
SL	Polystyrene-poly(L-leucine)
BCK	Polybutadiene-poly(carbobenzoxy-L-lysine)
SCK	Polystyrene-poly(carbobenzoxy-L-lysine)
BK	Polybutadiene-poly(L-lysine)
SK	Polystyrene-poly(L-lysine)
BE	Polybutadiene-poly(L-glutamic acid)
SE	Polystyrene-poly(L-glutamic acid)

I. Introduction

One of the main aims of macromolecular chemists is the synthesis of new materials whose properties are perfectly adapted to their utilization. Synthesis of new monomers cannot resolve all problems and composite materials have been the object of increasing development during the last 20 years. However, the formation of polymeric blends is generally prevented by the incompatibility of polymeric chains and it is difficult to prepare composite materials exhibiting all the desirable properties present in their components. A way of overcoming the inconveniences of the polymer incompatibility is the formation of covalent bonds between the constituents to obtain block or graft copolymers.

The possibility of employing block copolymers as materials that might possess desirable properties was originally considered by Mark[1]. In the first period the effort in preparing block copolymers was directed to radical polymerization[2–5] and it was only in 1956 that Szwarc obtained well-defined block copolymers by anionic polymerization[6]. In block copolymers, the incompatibility between polymeric chains becomes an advantage: a phase separation of the blocks occurs leading to the formation of microdomains which are responsible for the specific properties of block copolymers. For instance, the presence in a molecule of an elastomeric block linked by its ends to thermoplastic blocks generates a polymer in which reversible physical multifunctional cross-links allow the behaviour of conventional vulcanized elastomers at room temperature, but the material remains easily moldable at elevated temperature just as normal thermoplastic resins[7–10].

As block copolymers are still rather expensive materials, it may be advantageous to use them as additives to important industrial polymers. In this domain, possibilities are extremely numerous and diverse. They include an improvement of chemical properties such as resistence to degradation agents, or rheological properties such as adhesion of vinylic paints, high impact properties of conventional thermoplastics, or a compatibilization of polyolefins, polystyrene and poly(vinyl chloride) allowing the reuse of polymeric waste products. The above examples illustrate the great intrinsic potential of block copolymers in the quest of new materials with specific properties.

As the specific properties of block copolymers are related to the formation of microdomains, the present review will deal with the shape, size, and arrangements of microdomains in block copolymer systems. The phenomenon of phase separation and of microdomains formation is not restricted to dry copolymers, but is also a characteristic feature of their concentrated solutions, where mesophases are observed[11], and, therefore, we shall generally consider systems copolymer/solvent and treat dry copolymers as their limiting case.

After a brief recall of the methods of synthesis of block copolymers, we shall describe the principal types of organized structures, which have been observed in block copolymers both in mesomorphic and dry states. Then we shall examine the structure and properties of the most important block copolymers dividing them in three categories: copolymers with amorphous blocks, copolymers with amorphous and crystallizable blocks, copolymers with blocks presenting biological interest. However, in this review, only such properties will be taken into account that are related to the microdomain structure of block copolymers.

II. Principle of Block Copolymers Synthesis

Basically two methods lead to the formation of block copolymers. In the first one, the addition of a monomer B to a homopolymer of A which remains or becomes active and capable of initiating the polymerization of the monomer B gives a copolymer AB. In the second one, two independently prepared homopolymers A and B are linked by one of their ends to form a copolymer AB.

The preparation of block copolymers may be performed by a number of techniques[2,12], but the most suitable materials for the synthesis of block copolymers are living polymers which are generally obtained by anionic polymerization.

A) Anionic Polymerization

As all addition polymerizations, anionic polymerization proceeds in three steps: initiation, propagation and termination[13].

1) Initiation

It is the process forming the reactive centers from which macromolecules evolve. It may result from two different mechanisms: a nucleophilic attack of the monomer by an organometallic initiator (butyllithium, cumylpotassium, benzylsodium, etc.), or the transfer to the monomer of the counterion and the extra electron of an electron-transfer initiator (lithium or sodium naphtalene, biphenyl . . . .); later the ion-radical monomer having an extra electron in its lowest antibonding π orbital becomes a dicarbanion by dimerization of two activated monomer molecules. The use of a monofunctional initiator leads to a biblock copolymer AB, while that of a bifunctional initiator leads to a triblock copolymer ABA.

2) Propagation

The polymerization proceeds by a consecutive addition of monomeric units to reactive centers and after each addition cycle a new reactive center is regenerated at the end of the growing macromolecule. The propagation stops when all the monomer has been consumed. The polymerization proceeds again if a new amount of monomer is added. If one adds a second monomer B different from the first one A generally a block copolymer AB is obtained. The initiation of the polymerization of B by the living ends of A is only possible if the electroaffinity of B is higher or at least equal to that of A. When the two monomers have similar electroaffinities, the order of addition is indifferent.

Living polymers can only exist in aprotic solvents. They are killed by water, oxygen and a high number of electrophilic substances. Operating in absence of killing impurities one obtains stable living species; they are ionic species whose exact form (free ion, contact ion-pair, solvent separated ion-pair[14–17]) depends upon the concentration and the nature of the monomer, the counterion and the solvent; polar solvents favor the dissociation of ion-pairs while non polar mediums favor their association. The different species present during the polymerization participate in the growing chains according to their reactivity and some species can even have an influence on the stereo-regularity of the polymers formed.

3) Termination

Living species may be converted into dead ones by addition of a molecule with a mobile hydrogen atom (water, alcohol, etc.). Termination may also be performed by electrophilic molecules and in that case polymers are terminated by functional groups that can be used to initiate the polymerization of other monomers or as coupling adents between macromolecular chains.

The use of living anionic polymerization allows the synthesis of linear AB or ABA copolymers with blocks of predetermined size (in absence of transfer and termination reactions, the number-average molecular weight of one block is determined by the ratio of the mass of monomer and of the number of active centers) and the Poisson type molecular weight distribution is obtained if the following conditions are fullfilled: absence of termination or chain-transfer, presence of all initiating species at the onset of the reaction, perfectly uniform spatial monomer concentration, propagation rate constant independent of the molecular size, irreversible propagation. Conditions required for such anionic polymerization of many monomers have been reviewed by Fetters[18)].

B) Combination of Different Polymerization Methods

Not all monomers are anionically polymerizable. Nevertheless, one can take advantage of the activity of the living ends to introduce reactive end groups at the extremity of homopolymers and then use such end groups to initiate the polymerization of anionically non polymerizable monomers. This method has been applied to the synthesis of copolymers with polyvinyl and polylactone blocks[19)] and of copolymers with polyvinyl and polypeptide blocks[20–25)]. One can at last use both anionic and cationic polymerization to prepare block copolymers of tetrahydrofuran with styrene or methylstyrene[26)].

III. Organized Structures of Block Copolymers

A) Methods of Structural Characterization

The most profitable methods to study the block copolymers ordered structures are X-ray diffraction and electron microscopy. But differential scanning calorimetry, polarization microscopy, dilatometry, infrared spectroscopy and circular dichroism may bring supplementary information especially in the case of copolymers with crystallizable or polypeptide blocks[11].

X-ray diffraction and electron microscopy may be used separately or together. X-ray diffraction was applied at first in Strasbourg to systems copolymer/solvent[27–28] as a generalization of the study of systems soap/water and both experimental technics and structural interpretation methods[29,30] were extended to copolymers. Electron microscopy has been very popular in Japan. The japanese authors[31,32] have generally studied films obtained by evaporation of a dilute solution of copolymers; but, in that case, the structures observed may be influenced by the nature of the solvent and its speed of evaporation[33,34]. Low angle X-ray diffraction and electron microscopy have been systematically applied together to systems copolymer/solvent and to dry copolymers in Orleans[11,35] and to oriented dry copolymers in Bristol[36].

The respective advantages and disadvantages of X-ray diffraction and electron microscopy have been analyzed in detail in two recent reviews[11,36] so we prefer to describe here the method of structural determination which, according to the author's experience, is the safest one. This method combines the use of both X-rays and electron microscopy. It might seem that the presence of a solvent would prevent electron microscopy from being applied. The use of a monomer as a solvent and its postpolymerization allow the utilization of electron microscopy for the study of copolymer/solvent systems as well as for the study of dry copolymers[11,35]. In Orleans we proceed as follows[37]: we dissolve the copolymer in a monomer which is a preferential solvent of one block and we obtain a mesomorphic gel; we determine the structure of the gel by low-angle X-ray diffraction; we polymerize the solvent by UV light or with a peroxide and we obtain a solid organized copolymer; we verify by low angle X-ray diffraction that the periodic structure has not been destroyed by the polymerization of the monomer and we measure its new parameters; then we cut the solid sample with an ultramicrotome, stain one of its block (generally a diene block) with osmium tetroxide and observe the structure by electron microscopy.

With this method we have always obtained excellent results. Tables 1, 2, 3 illustrate the agreement between values of structural parameters determined by X-rays and electron microscopy.

B) Description of the Periodic Structures

In copolymer ordered structures, the segregated microphases can be spheres, cylinders or lamellae. The lamellae tend to form a regularly repeating lamellar sequence, the cylinders arrange themselves in a bidimensional hexagonal lattice, while the spheres give rise to cubic lattices.

We shall briefly describe these structures and recall how they are characterized by X-rays and electron microscopy.

1) Lamellar Structure

The lamellar structure is characterized by low-angle X-ray patterns exhibiting a set of sharp lines with Bragg spacings in the ratios 1, 2, 3, 4, 5 . . . . and by micro-

Table 1. Examples of structural parameters (in Å) of organized copolymers with a lamellar structure. Low Ange X-ray diffraction values (XR), electron microscopy values (EM)
d Total thickness of a sheet.
d_A Thickness of the non-polydiene layer.
d_B Thickness of the polydiene layer

Copolymer	% Of polymerized solvent	d		d_A		d_B	
		XR	EM	XR	EM	XR	EM
SB 34	25 MMA	446	460	254	250	192	210
BSB 375	20 Styr	466	450	292	280	174	170
SBS 332	27 AcOV	452	440	262	260	190	180
BMS 31	23 Styr	288	300	156	160	132	140
BVN 41	35 MMA	335	325	223	215	112	100
IVP 12	21 AcOV	754	730	397	370	357	360
BG 530	diClpropene	260	240	200	185	60	55

Table 2. Example of structural parameters (in Å) of organized copolymers with a hexagonal structure. Low-angle X-ray diffraction values (XR); electron microscopy values (EM)
D Distance between the axis of two neighbouring cylinders.
2 R Diameter of the cylinders filled by the insoluble polybutadiene blocks

Copolymer	% Polymerized solvent	D		2 R	
		XR	EM	XR	EM
SB 31	25 Styr	322	335	141	150
BSB 372	25 MMA	358	370	163	175
SBS 333	30 MMA	476	460	280	265
BMS 41	24 Styr	356	345	78	70

Table 3. Example of structural parameters (in Å) of organized copolymers with an inversed hexagonal structure. Low-angle X-ray diffraction values (XR), electron microscopy values (EM)
D Distance between the axis of two neighbouring cylinders.
2 R Diameter of cylinders filled by the non-polydiene blocks swollen by the solvent

Copolymer	% Of polymerized solvent	D		2 R	
		XR	EM	XR	EM
SB 35	33 MMA	576	560	443	450
BSB 421	29 Styr	425	435	285	292
SBS 365	25 MMA	563	575	398	410
BVN 11	32 MMA	382	370	145	140
IVP 22	dry	205	200	96	86

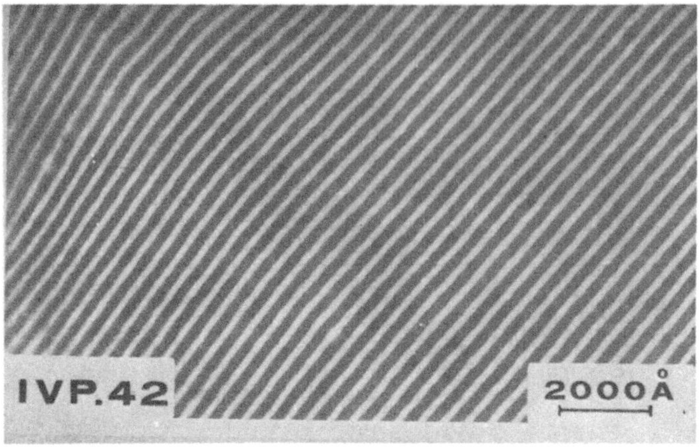

Fig. 1. Example of electron micrograph of the lamellar structure. Copolymer polyisoprene-poly(vinyl-2-pyridine) (IVP.42) containing 60,5% of polyisoprene, swollen with 25% MMA and post-polymerized.
white stripes: poly(vinyl-2-pyridine) layer; black stripes: polyisoprene layer stained by osmium tetroxide

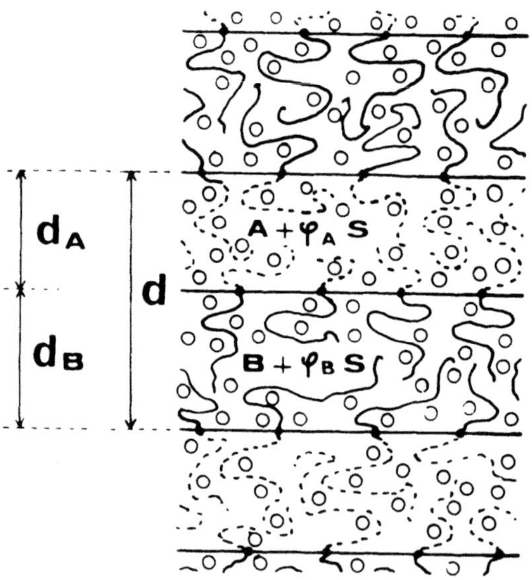

Fig. 2. Schematic representation of the lamellar structure in the case of a non-selective solvent.
For a selective solvent $\varphi_A = 1$ and $\varphi_B = 0$ and the solvent is localized in the A layer

graphs presenting alternating dark and light parallel stripes (Fig. 1); the black stripes contain the stained blocks (generally polydiene) and the white stripes contain the other blocks.

Therefore, the lamellar structure (sketched in Fig. 2) is formed by planar, parallel equidistant sheets; each sheet results from the superposition of two layers one formed by the A blocks and the other by the B blocks. The solvent may be absorbed by only one layer or it may exist a partition of the solvent between the two layers.

The characteristic parameters of the lamellar structure are the total thickness d of a sheet, the thickness d_A of the layer containing the most soluble blocks, the thickness d_B of the layer containing the less soluble blocks, and the average surface Σ available for a molecule at the interface.

The total thickness $d = d_A + d_B$ of a sheet is directly given by the Bragg spacing of the X-ray patterns, or directly measured on the electron micrographs (to obtain an accurate value of d by electron microscopy one must use electron micrographs provided by sections perpendicular to the planes of the sheets, this is easy with an electron microscope equipped with a goniometer head).

The thickness d_A and d_B of the two layers are directly measured on the micrographs or calculated from the intersheet spacing d given by X-rays using a formula based on simple geometrical considerations

$$d_B = d \left[1 + \frac{CX_A v_A + (1-C)\varphi_A v_S}{CX_B v_B + (1-C)\varphi_B v_S}\right]^{-1} \quad (1)$$

$$\Sigma = \frac{2 M_B v_B}{N d_B}\left[1 + \frac{X_A}{X_B}\cdot\frac{v_A}{v_B} + \frac{(1-C)}{C}\cdot\frac{v_S}{v_B}\cdot\frac{1}{X_B}\right] \quad (2)$$

where C is the polymer concentration in solution, X_B concentration of the B block in the copolymer, v_A and v_B are specific volumes of the A and B blocks, v_S is the specific volume of the solvent, φ_A and φ_B are partition coefficients of the solvent ($\varphi_A + \varphi_B = 1$), M_B is the number-average molecular weight of the B block, and N the Avogrado number.

If the solvent dissolves only one block (A for instance), $\varphi_A = 1$ and $\varphi_B = 0$, then:

$$d_B = d\left[1 + \frac{CX_A v_A + (1-C) v_S}{CX_B v_B}\right]^{-1} \quad (3)$$

$$\Sigma = \frac{2 M_B v_B}{N d_B} \quad (4)$$

2) Cylindrical Hexagonal Structures

Cylindrical hexagonal structures exhibit low-angle X-ray patterns characterized by a set of sharp lines with Bragg spacings in the ratio $1, \sqrt{3}, \sqrt{4}, \sqrt{7}, \sqrt{9}, \ldots$. To characterize these structures by electron microscopy, one needs micrographs corresponding to sections of the sample in two perpendicular directions (Figs. 3 and 4). The sections perpendicular to the direction of the axis of the cylinders are characterized by circular spots arranged in a hexagonal array (main Figs. 3 and 4). The sections parallel to the direction of the cylinder axis are characterized by alternating dark and light parallel stripes (inserts of Figs. 3 and 4).

X-ray diffraction cannot distinguish the different blocks and is not able to say which types of blocks form the cylinders and the matrix. On the contrary,

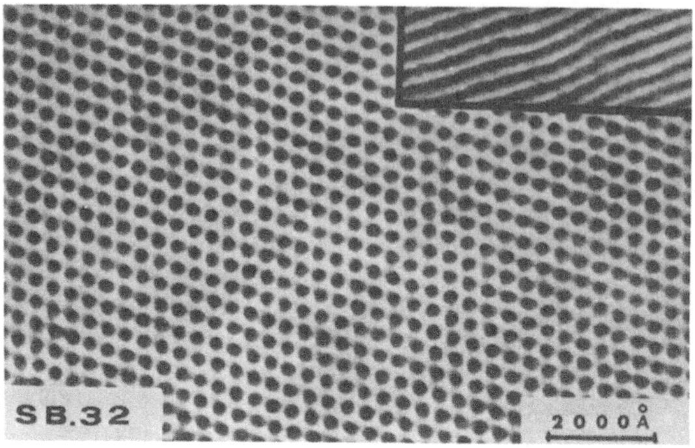

Fig. 3. Example of electron micrograph of a hexagonal structure. Copolymer polystyrene-polybutadiene SB. 32 containing 30,5% of polybutadiene, swollen with 29% of MMA and post-polymerized. Main figure: section along the plane perpendicular to the direction of the axis of the insoluble polybutadiene cylinders; insert: section by a plane parallel to the axis of the cylinders. Polybutadiene stained by osmium tetroxide in dark

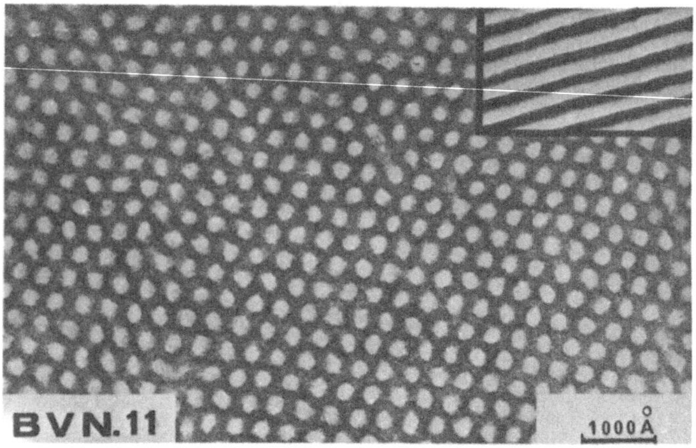

Fig. 4. Example of electron micrograph of the inverse hexagonal structure. Copolymer polybutadiene-poly(vinyl-2-naphthalene) BVN. 11 containing 62% polybutadiene, swollen with 36% MMA, and post-polymerized[87]. Main figure: section along the plane perpendicular to the direction of the axis of the poly(vinylnaphthalene) cylinders swollen with the solvent; insert: section by a plane parallel to the axis of the cylinders. Polybutadiene stained by osmium tetroxide in dark

electron microscopy easily visualizes the two types of blocks if one of them is stained. Furthermore, as it is illustrated by the comparison of Figs. 3 and 4, electron microscopy demonstrates the existence of two types of cylindrical hexagonal structures: the hexagonal (Fig. 3) and the inverse hexagonal (Fig. 4). In the hexagonal structures (Fig. 3), the cylinders are filled with insoluble blocks (generally the diene blocks: the spots are black) and the matrix is formed by the soluble blocks. In the inverse hexagonal structure (Fig. 4), the cylinders are filled with the soluble blocks

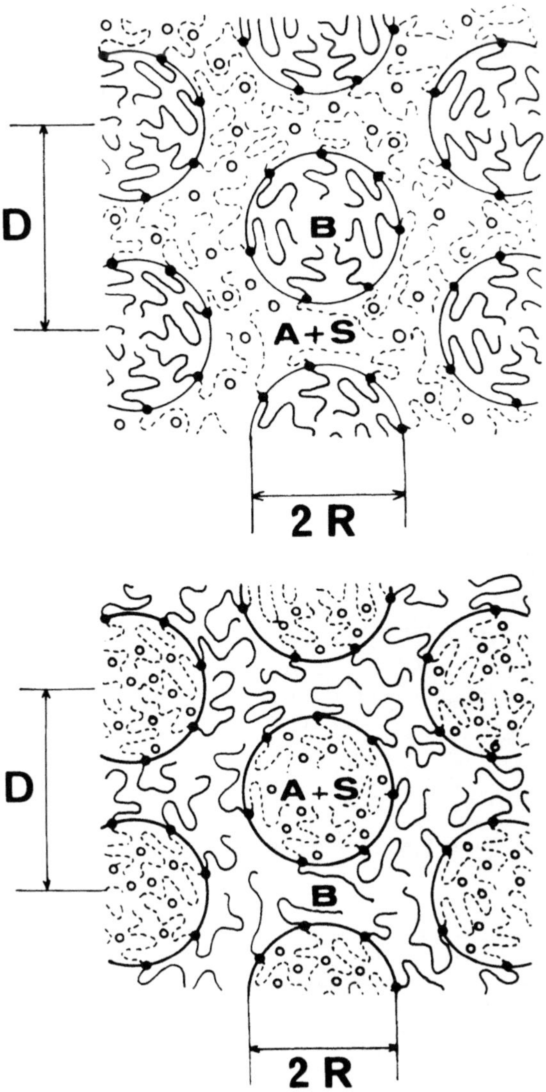

5a

5b

Fig. 5. Schematic representation of the hexagonal and inverse hexagonal structures.
a hexagonal structure, the solvent swells the matrix, b inverse hexagonal structure, the solvent swells the cylinders

and the matrix is formed by the insoluble blocks (generally the diene blocks: the matrix is black). In both the hexagonal and the inverse hexagonal structure, the cylinders are filled with the shortest blocks and the matrix is formed by the largest ones[11, 37].

The characteristic parameters of both the hexagonal and the inverse hexagonal structures (Fig. 5) are the distance D between the axis of two neighbouring cylinders, the diameter 2 R of the cylinders, and the specific area available for a molecule at the interface.

D is directly measured on electron micrographs or directly obtained from the Bragg spacings of X-ray patterns, 2 R is directly measured on electron micrographs or calculated from the value of D given by X-rays using the following formulae:

$$R^2 = \frac{D^2\sqrt{3}}{2\,\pi}\left[1 + \frac{CX_m v_m + (1 - C)\,\varphi_m v_S}{CX_i v_i + (1 - C)\,\varphi_i v_S}\right]^{-1} \tag{5}$$

$$\Sigma = \frac{2\,M_i v_i}{N\,D}\cdot\frac{2\,\pi}{\sqrt{3}}\left[1 + \frac{X_m}{X_i}\,\frac{v_m}{v_i} + \frac{1 - C}{C}\,\frac{v_S}{v_i}\right]\left[1 + \frac{1 - C}{C}\,\frac{v_S}{v_i}\,\frac{\varphi_i}{X_i}\right]^{1/2} \tag{6}$$

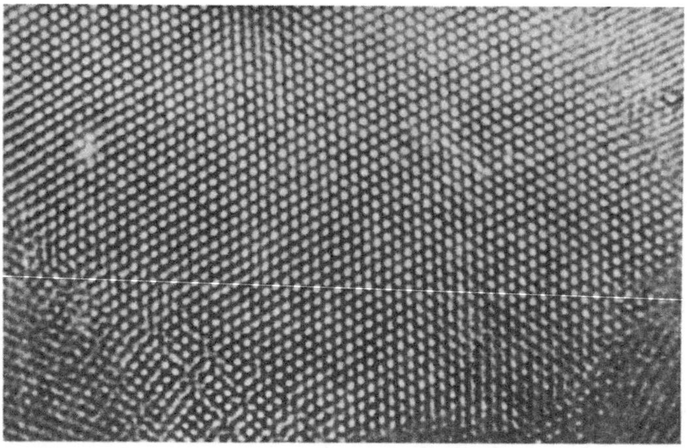

6a

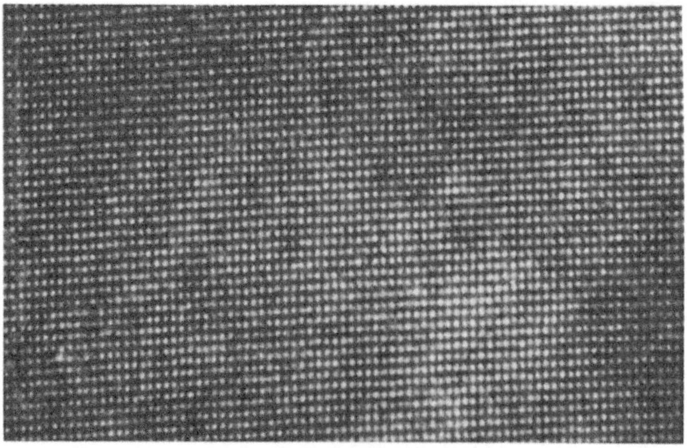

6b

Fig. 6. Examples of electron micrographs of the inverse centered cubic structure[38]. Copolymer polystyrene-polyisoprene-polystyrene SIS.1107 containing 10% polystyrene. White circles are polystyrene spheres.
a section by a plane 111 (hexagonal arrangement), b section by a plane 110 (rectangular arrangement)

where C is the polymer concentration in solution, X_i the concentration in the copolymer of the block located inside the cylinders, v_i and v_m are specific volumes of the blocks located inside the cylinders and in the matrix, v_S is the specific volume of the solvent, φ_i and φ_m are partition coefficient of the solvent between the cylinders and the matrix ($\varphi_i + \varphi_m = 1$), and M_i is the number-average molecular weight of the blocks located inside the cylinders; X_m is the concentration in the copolymer of the block forming the matrix.

3) Cubic and Inverse Cubic Structures

Cubic structures can be of three types: simple, body centered or face centered. Their low angle X-ray patterns are characterized by the following sequences of reflexions when the diffraction angle increases:
– $1, \sqrt{3}/\sqrt{4}, \sqrt{3}/\sqrt{8}, \sqrt{3}/\sqrt{11}, \ldots$ for a face centered cubic lattice
– $1, 1/\sqrt{2}, 1/\sqrt{3}, 1/\sqrt{4}, \ldots$ for a simple cubic lattice
– $1, \sqrt{2}/\sqrt{4}, \sqrt{2}/\sqrt{6}, \sqrt{2}/\sqrt{8}, \ldots$ for a body centered cubic lattice

Until now, only the existence of a body centered cubic lattice has been clearly demonstrated[38].

Electron micrographs of the centered cubic lattice depends upon the plane of the section: one obtains circular spots arranged on a hexagonal lattice (plane 111), on a square lattice (plane 100) or on a rectangular lattice (plane 110). Figures 6 a and 6 b illustrates this effect for a copolymer SIS containing 10% of polystyrene studied by Italian authors[38].

Electron microscopy has also allowed to show the existence of two types of cubic structures: the body centered cubic structure characterized (in the case of copolymers containing polydiene) by black spots on a white background (Fig. 7)

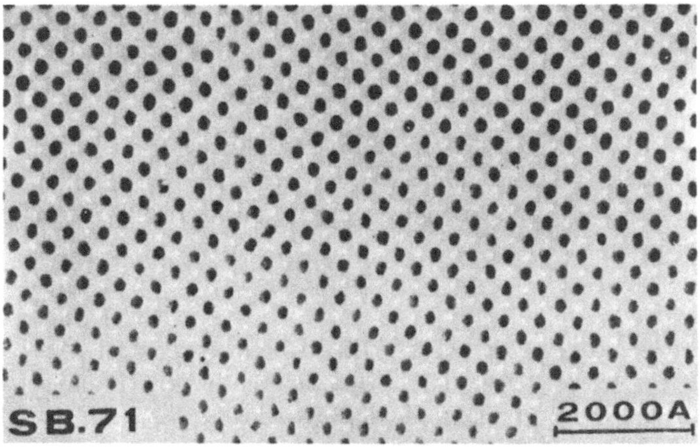

Fig. 7. Example of electron micrograph of the centered cubic structure. Copolymer polystyrene-polybutadiene SB.71 containing 13% polybutadiene, swollen with 25% styrene, and post-polymerized. Black circles are polybutadiene spheres

and the inverse body centered cubic structure characterized by white spots on a black background (Figs. 6 a and 6 b).

The characteristic parameters of both the body centered cubic structure and the inverse body centered cubic structure are the side a of the cell, the diameter $2R$ of the spheres, and the specific surface Σ.

The side a is directly measured on electron micrographs or directly provided by the Bragg spacings of X-ray patterns. $2R$ is directly measured on electron micrographs or calculated from the value of a given by X-rays.

$$R_{cc}^3 = \frac{3a^3}{8\pi}\left[1 + \frac{CX_m v_m + (1-C)\varphi_m v_S}{CX_i v_i + (1-C)\varphi_i v_S}\right]^{-1} \tag{7}$$

$$\Sigma = \frac{3M_i}{Na}\left(\frac{8\pi}{3}\right)^{1/3}\left[v_i + \frac{1-C}{C} v_S \frac{\varphi_i}{X_i}\right]\left[1 + \frac{CX_m v_m + (1-C)\varphi_m v_S}{CX_i v_i + (1-C)\varphi_i v_S}\right]^{1/3} \tag{8}$$

IV. Theories of Microdomain Formation

We shall recall the principal theories dealing with the microphase separation and the morphology of block copolymers, but very briefly as they have been recently reviewed in details by Folkes and Keller[36].

Dale Meier has been one of the firsts if not the first who presented a theory of domain formation in block copolymers[39]. In its original version[39], Meier's theory was restricted to AB block copolymers and spherical domains. In a series of following papers[40–45], however, Meier has refined his theory considering different shapes of domains, the effect of the presence of a solvent, the dimensions of the interface, the interfacial properties of block copolymers and the solubilization of homopolymers by copolymers.

S. Krause[46, 47] has studied the general conditions of phase segregation and found (in agreement with Meier's results) that the phase separation is more difficult for block copolymers than for homopolymers and that this difficulty increases with the number of blocks.

Bianchi *et al.*[48, 49] have tried to predict the equilibrium dimensions of microdomains assumed as spherical.

Leary and Williams have proposed for ABA block copolymers a model characterized by a simultaneous existence of three phases: pure A, pure B and a mixed region. Their model, which postulated spherical domains in its first version[50], has been extended to cylindrical and lamellar domains[51, 52].

Krigbaum has formulated a theory predicting the dimensions of the domains in the case of a lamellar structure. For AB copolymers[53], each subchain behaves as a random flight with reflecting barriers. For BAB copolymers[54], the possibility of existence of bridge and loop states for the central A block is taken into account and the effects of the introduction of reflecting and neutral barriers are compared.

Helfand has presented a general theory of inhomogeneous polymeric systems[55] and has applied it to the interface between immiscible homopolymers[56, 57] and to block copolymers[58, 59]. In order to simplify the problem of the numerical calculation of the microdomain sizes, he has derived from his general theory a narrow interface approximation[60]. The values predicted by the simplified theory for the dimensions of the lamellar block copolymer microdomains are in good agreement with the experimental results[60, 61].

The theoretical consideration developed by Krämer, Hofman and Kämpf[62] and by Kawai *et al.*[63, 64] to explain the behaviour of their copolymers will be dealt with together with the respective experimental results.

V. Copolymers with Amorphous Blocks

A wide range of copolymers with two or three amorphous blocks have been synthetized and studied. To report the principal results concerning their structure and their properties related to their structure we shall divide the copolymers into three classes according to the chemical nature of the blocks. We shall consider successively copolymers containing polybutadiene blocks, copolymers containing polyisoprene blocks, and copolymers containing only nonpolydiene blocks. The order adopted for this classification reflects the amount of studies and the importance of results obtained with the three categories of copolymers. For each class of copolymers, we shall review at first the results of the studies performed on systems copolymer/solvent, that is to say, on the block copolymer mesophases, and then on copolymers in bulk, dry copolymers being limiting cases of their concentrated solutions[65].

A) Block Copolymers of Butadiene and Other Monomers

We shall examine successively the behaviour of AB copolymers of butadiene and styrene, butadiene and α-methylstyrene, butadiene and vinylnaphtalene and of BAB and ABA copolymers of butadiene and styrene.

1) Mesophases of Polystyrene-polybutadiene Block Copolymers (SB)

Douy has synthetized polystyrene-polybutadiene (SB) block copolymers of various molecular weights and compositions[66] by anionic polymerization, under high vacuum, in tetrahydrofuran dilute solution (less than 5%), at low temperature (–70 °C), and with cumyl potassium as initiator. Resulting from the polymerization conditions, the microstructure of the polybutadiene block is 90% 1,2 and 10% 1,4.

a) Structure and Range of Stability of the Mesophases

Low-angle X-ray diffraction and electron microscopy have demostrated that SB copolymers exhibit mesophases in different solvents for solvents concentrations smaller than about 45% and that the periodic structure of the mesophases in neither destroyed by slow evaporation of the solvent, nor by polymerization of the solvent[66, 71].

The type of structure adopted by the copolymer is governed by its composition. When the polybutadiene content of the copolymer increases from 10 to 90%, one observes successively:

a centered cubic structure formed by spheres of polybutadiene in a polystyrene matrix, for a polybutadiene content between 10 and 16%;
an orthorhombique structure formed by short rods of polybutadiene in a polystyrene matrix, for a polybutadiene content between 16 and 18%[72];
a hexagonal structure formed by cylinders of polybutadiene in a polystyrene matrix for a polybutadiene content between about 18 and 36%;
a lamellar structure formed by alternating layers of polybutadiene and polystyrene for a polybutadiene content between about 36 and 60%;
an inverse hexagonal structure formed by cylinders of polystyrene in a polybutadiene matrix for a polybutadiene content between about 60 and 80%;
an inversed centered cubic structure formed by spheres of polystyrene embedded into a polybutadiene matrix for polybutadiene content higher than about 80%.

The range of stability of the mesophases has been studied by differential scanning calorimetry (DSC) and low-angle X-ray diffraction. DSC has enabled the determination of phase transitions and ranges of stability of pure phases, X-ray diffraction has given information about the structure of the different phases and the variation of the structural parameters with temperature[73]. It has been found that SB copolymers exhibit only one mesophase as a function of temperature and concentration, the structural type of the mesophase being determined by the composition of the copolymer. The thermal stability of the mesophase is increased by annealing at solvent concentration smaller than about 20% and by polymerization of the solvent at solvent concentration between about 20 and 40%[74]. Examples of temperature/concentration phase diagrams are given in Figs. 8 and 9 for the hexagonal and lamellar structures.

b) Factors Governing the Structural Parameters

The principal factors governing the geometrical parameters of the periodic structures of the mesophases are the concentration, nature and polymerization of the solvent, temperature, lengths of the blocks and total molecular weight of the copolymer.

b 1) Influence of the Concentration of the Solvent. SB copolymers have been studied in solution in methylethyl ketone (MEK), methyl methacrylate (MMA), styrene (Styr), vinyl acetate (AcoV) and ethyl acetate (AcoEt) which are all preferential solvents of the polystyrene blocks[66]. The selectivity of these solvents has been demonstrated by DSC (study of the glass transitions of the two blocks)[35],

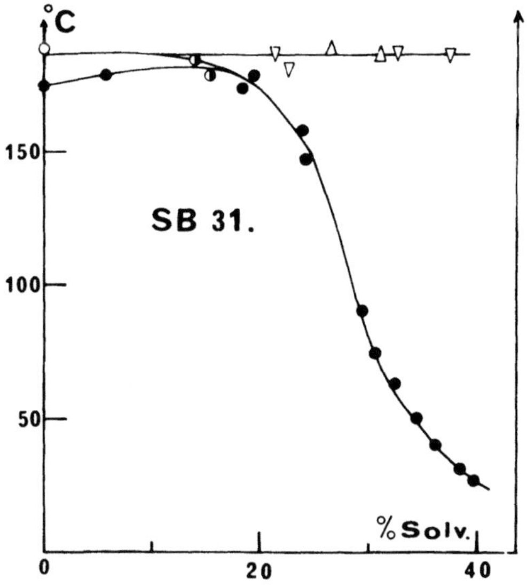

Fig. 8. Example of phase diagram concentration/temperature. Copolymer polystyrene-polybutadiene SB. 31 containing 20% polybutadiene and exhibiting a hexagonal structure.
•: MEK; ○: MEK annealed; ▿: p. styr.; ▵: p. MMA

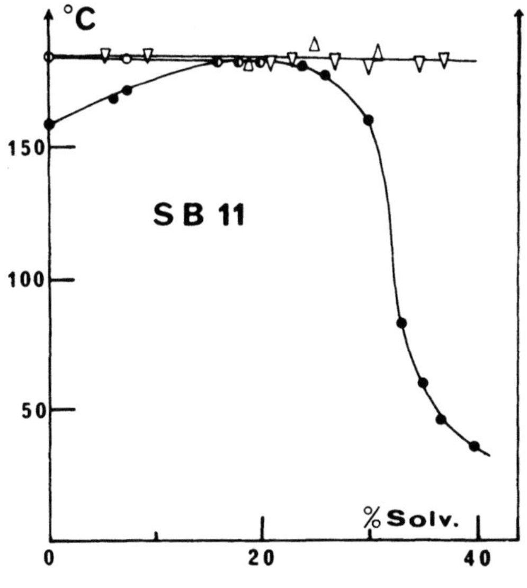

Fig. 9. Example of phase diagram concentration/temperature. Copolymer polystyrene-polybutadiene SB. 11 containing 39% polybutadiene and exhibiting a lamellar structure.
•: MEK; ○: MEK annealed; ▿: p. styr.; ▵: p. MMA

X-ray diffraction (study of the respective intensities of the different diffraction orders)[75], and electron microscopy[11].

It has been shown that when the solvent concentration increases:
the lattice parameters: total thickness d of a sheet for the lamellar structure (Fig. 10), distance *D* between the axis of two neighbouring cylinders for the hexagonal structure (Fig. 12) and the inverse hexagonal structure (Fig. 13), the

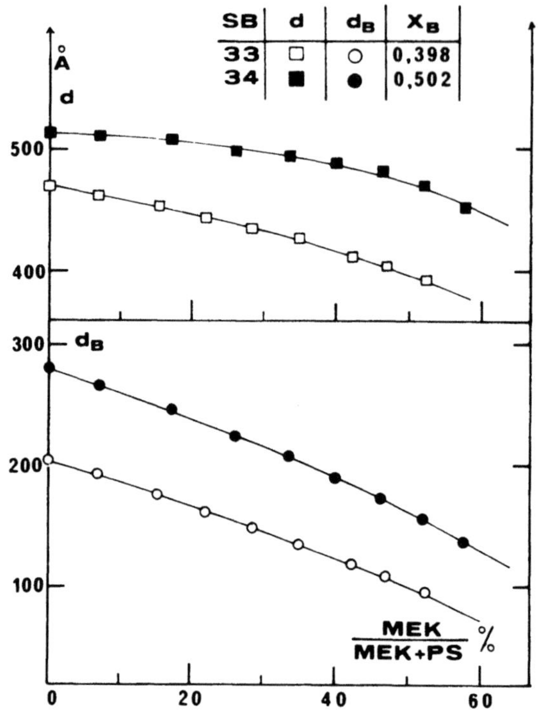

Fig. 10. Variation with the swelling ratio of the polystyrene block of the total sheet thickness d and of the thickness d_B of the insoluble layer of the copolymers SB. 33 and SB. 34 containing, respectively, 39.8% and 50.2% polybutadiene. SB.33: □: d, ○: d_B; SB.34: ■: d, ●: d_B

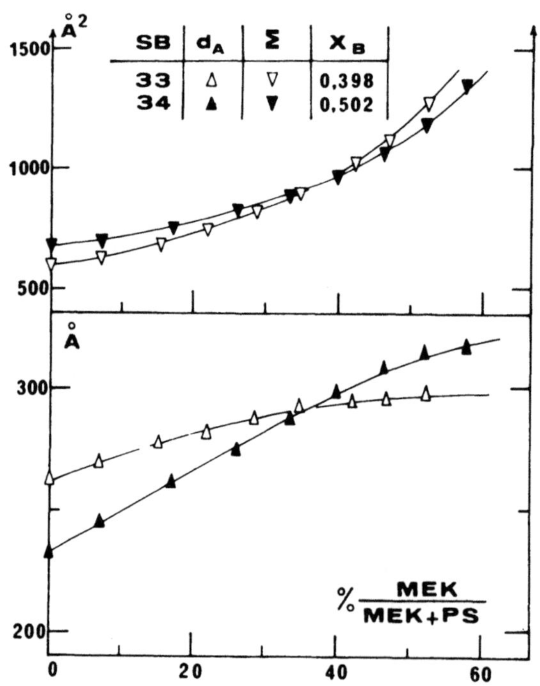

Fig. 11. Variation with the swelling ratio of the polystyrene block of the thickness d_A of polystyrene soluble layer and the specific surface Σ for the copolymers SB.33 and SB.34. SB.33: Δ: d_A, Δ: Σ; SB.34: Δ: d_A, Δ: Σ

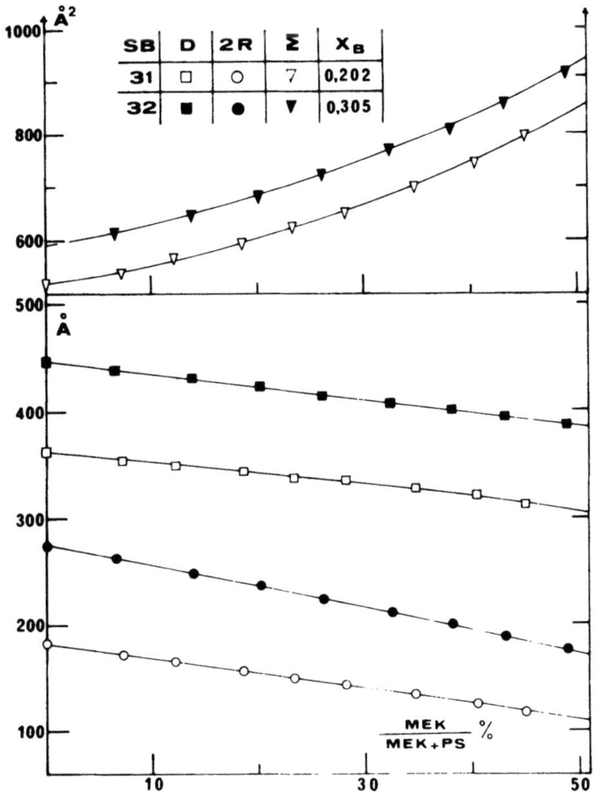

Fig. 12. Variation with the swelling ratio of the polystyrene block of the geometrical parameters of the hexagonal structure. D distance between the axis of two neighbouring cylinders, 2 R diameter of the polybutadiene cylinders, Σ specific surface. Copolymers SB.31 and SB.32 containing, respectively, 20.2% and 30.5% polybutadiene.
SB.31: □: D, ○: 2 R; ▽: Σ; SB.32: ■: D, ●: 2 R, ▼: Σ

side *a* of the cell for the centered cubic structure (Fig. 14) and the inverse centered cubic structure (not represented) always decrease;

the specific surface Σ increases for the five types of structures (see Figs. 11 to 14);

the characteristic parameters of the domains containing the insoluble blocks: thickness d_B of the polybutadiene layer in the case of the lamellar structure (Fig. 10), the diameter 2 R of the polybutadiene cylinders in the case of the hexagonal structure (Fig. 12), and the diameter 2 R of the polybutadiene spheres in the case of the centered cubic structure (Fig. 14) always decrease;

the characteristic parameters of the domains containing the soluble blocks: thickness d_A of the polystyrene layer for the lamellar structure (Fig. 11), the diameter 2 R of the polystyrene cylinders for the inverse hexagonal structure (Fig. 13), and the diameter 2 R of the polystyrene spheres for the inverse centered cubic structure (not represented) always increase.

b 2) Influence of the Nature of the Solvent. The structure of SB copolymers has been exhaustively studied in three preferential solvents of the polystyrene blocks:

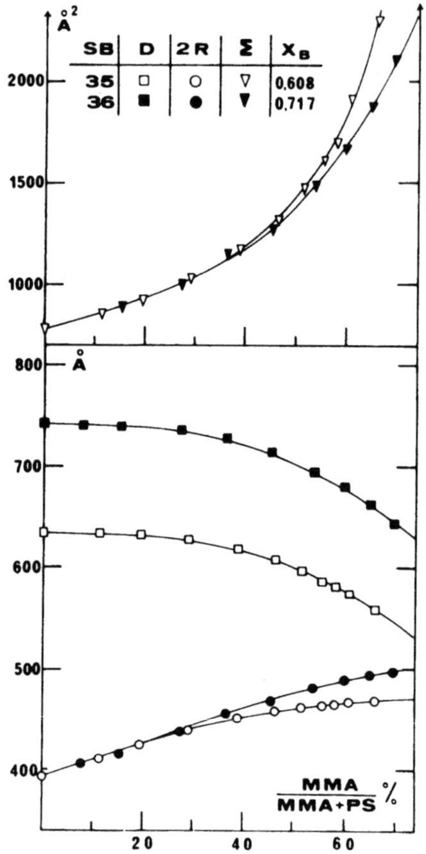

Fig. 13. Variation with the swelling ratio of the polystyrene block of the geometrical parameters of the inverse hexagonal structure.
D distance between the axis of two neighbouring cylinders, 2 R diameter of the polystyrene cylinders, Σ specific surface. Copolymers SB. 35 and SB. 36 containing respectively 60.8% and 71.7% polybutadiene.
SB. 35: □: D, ○: 2 R, ▽: Σ; SB. 36: ■: D, ●: 2 R, ▼: Σ

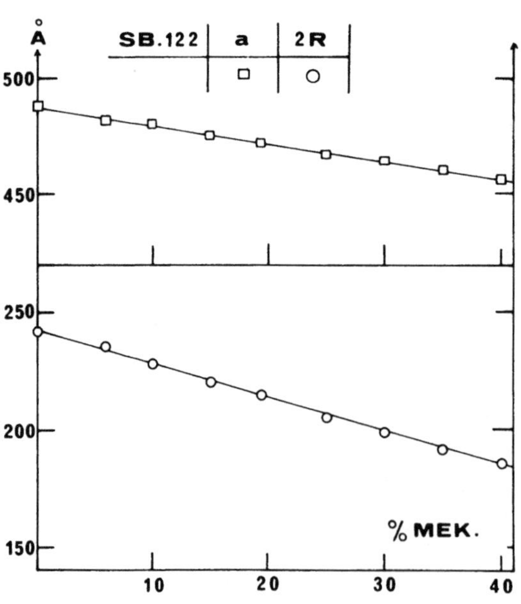

Fig. 14. Variation with solvent concentration of the geometrical parameters of the body centered cubic structure. Copolymer SB. 122 containing 11% polybutadiene in MEK solution.
□: a side of the cubic cell; ○: 2 R diameter of the polybutadiene cylinders

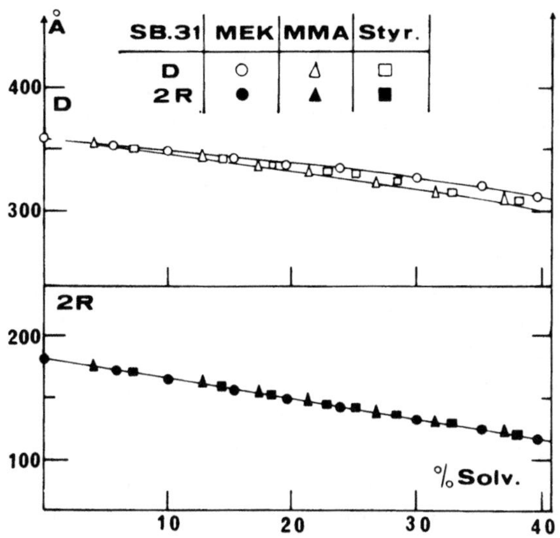

Fig. 15. Influence of the nature of the solvent on the parameters D and 2 R of the hexagonal structure for the copolymer SB. 31.
D: ○: MEK; △: MMA; □: Styr.; 2 R: ●: MEK; ▲: MMA; ■: Styr.

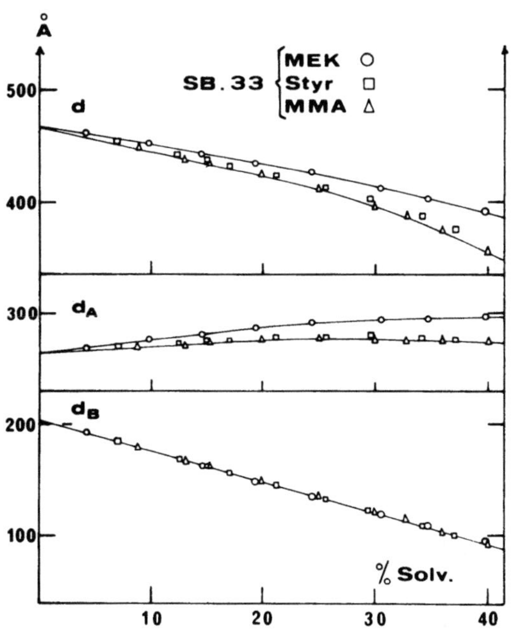

Fig. 16. Influence of the nature of the solvent on the parameters d, d_A and d_B of the lamellar structure for the copolymer SB. 33.
○: MEK; △: MMA; □: Styr.

methylethyl ketone, methyl methacrylate and styrene and, with less details, in other solvents such as vinyl acetate and ethyl acetate[66]. It has been shown that if the solvent remains a preferential solvent of the polystyrene blocks, its nature does not influence the characteristic parameters of the domains containing the insoluble polybutadiene blocks: thickness of the polybutadiene layer for the lamellar structure, diameter of the polybutadiene cylinders or spheres for the hexagonal and the centered cubic structures. The characteristic parameters of the soluble domains vary slightly with the nature of the solvent and their variation is correlated with the densities of the solvents. Figures 15 and 16 illustrate this behaviour in the case of the hexagonal and the lamellar structures, respectively.

b 3) Influence of the Polymerization of the Solvent. Mesomorphic gels prepared by swelling of the copolymer in a monomer (styrene, methyl methacrylate, vinyl acetate . . . .) have been converted into solid organized copolymers by polymerization of the monomer by UV light or by a peroxide[71]. Low-angle X-ray diffraction has shown that the structural type is not modified by the polymerization of the solvent and that some geometrical parameters change while some others remain constant. One observes an absence of variation of the characteristic parameters of the insoluble polybutadiene domains: diameter 2 R of the spheres for the centered cubic structure, diameter 2 R of the cylinders for the hexagonal structure and thickness d_B of the insoluble layer for the lamellar structure. On the contrary, the polymerization of the solvent is accompanied by a contraction of the characteristic parameters of the domains swollen by the monomer: diameter of the polystyrene spheres and cylinders of the inverse centered cubic structure and the inverse hexagonal structure, thickness d_A of the polystyrene layer of the lamellar structure, volume V_A of the polystyrene chains for the centered cubic and the hexagonal structures. Figures 17 to 19 illustrate these phenomena for the hexagonal, lamellar, and inverse hexagonal structures.

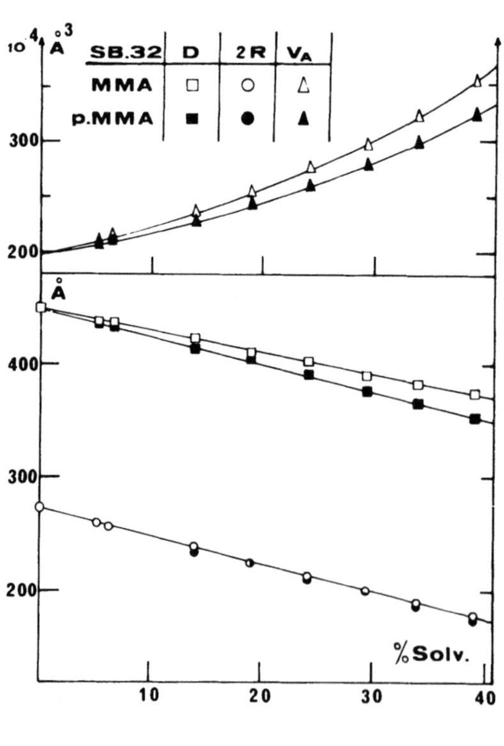

Fig. 17. Variation of the parameters D, 2 R and V_A of the hexagonal structure during polymerization of the solvent for the copolymer SB. 32 in solution in MMA.
V_A: average volume of a polystyrene

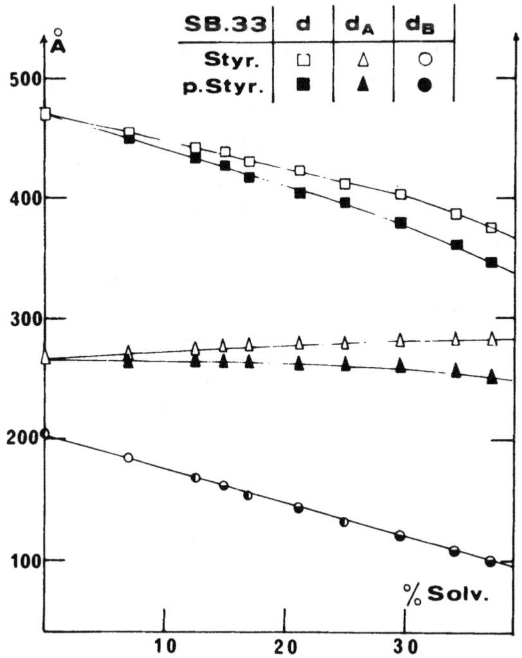

Fig. 18. Variation of the parameters d, d_A and d_B of the lamellar structure during polymerization of the solvent for the copolymer SB. 33 in solution in styrene

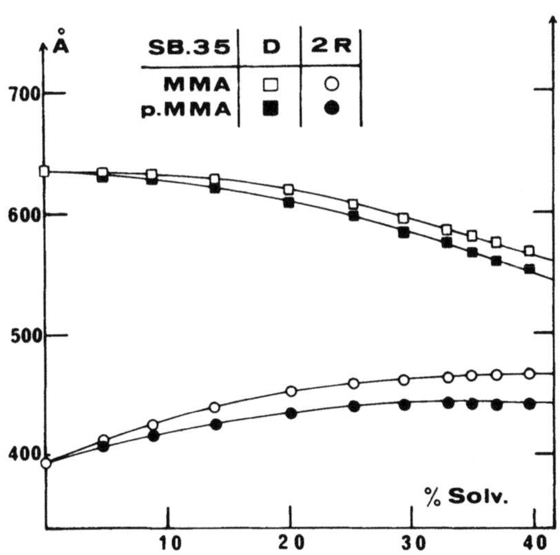

Fig. 19. Variation of the parameters D and 2 R of the inverse hexagonal structure during the polymerization of the solvent for the copolymer SB. 35 in solution in MMA

b 4) Influence of the Temperature. Low-angle X-ray diffraction has shown that with increasing temperature, the specific surface Σ increases while all the other parameters decrease[73].

b 5) Influence of the Molecular Weight of the Polybutadiene Blocks. In order to establish the influence of the molecular weight of the insoluble blocks on the parameters of different structures adopted by SB copolymers, a set of six copolymers (SB 31 to SB 36) with the same polystyrene block (M_A = 49000) but of different composition has been studied by low-angle X-ray diffraction and electron microscopy[66]. In this way, the variation of the structural parameters (within a fixed structural type) with the polybutadiene content of the copolymer or (which is equivalent) the molecular weight M_B of the polybutadiene block has been followed for three types of structures: lamellar, hexagonal, and inverse hexagonal. The ranges of stability in composition of the centered cubic and the inverse centered cubic structures are too small to allow such type of study.

In the case of the lamellar structure, the increase of the polybutadiene content X_B of the copolymer and of the molecular weight M_B of the polybutadiene block entails an increase of the total thickness d of a sheet and of the thickness d_B of the polybutadiene layer; Fig. 10 illustrates this behaviour for copolymer SB 33 and SB 34 which contain, respectively, 39.8% and 50.2% polybutadiene. The values of the thickness d_A of the polystyrene layer and of the specific surface Σ are not independent of the polybutadiene content of the copolymer as it is illustrated by Fig. 11.

In the case of the hexagonal structure the increase of the molecular weight M_B of the polybutadiene block is accompanied by an increase of both the distance D between the axis of two neighbouring cylinders and the diameter 2 R of the polybutadiene cylinders. Figure 12 illustrates these results in the case of copolymers SB 31 and SB 32 whose respective molecular weights of the polybutadiene blocks are 12400 and 21 500.

In the case of the inverse hexagonal structure, the distance D between the axis of two neighbouring cylinders increases with the molecular weight of the polybutadiene block and the swelling of the polystyrene cylinders increases with the polybutadiene content of the copolymer. Figure 13 illustrates this behaviour for copolymers SB 35 and SB 36 with polybutadiene molecular weight of 76000 and 124000, respectively.

b 6) Influence of the Molecular Weight of the Copolymer. It has been shown[76] that the lattice parameters total thickness *d* of a sheet for the lamellar structure, distance *D* between the axis of two neighbouring cylinders for the hexagonal and the inverse hexagonal structure increase monotonously with the molecular weight of the copolymer.

c) Desorganization of the Mesophases by Dilution

The desorganization by dilution of the periodic structures of block copolymers has been studied by electron microscopy after polymerization of the solvent[66, 70, 71]. Two types of solvents have been used: styrene, which is the monomer for the soluble polystyrene block and which prevents any incompatibility between polymeric chains during polymerization of the monomer, and methyl methacrylate which allows the study of the effect of incompatibility between polymeric chains during polymerization of the solvent.

c 1) Case of the Monomer of one Block. The progressive evolution with dilution of the systems copolymer polystyrene-polybutadiene/styrene monomer has been studied for the three following structures: lamellar, hexagonal and inverse hexagonal.

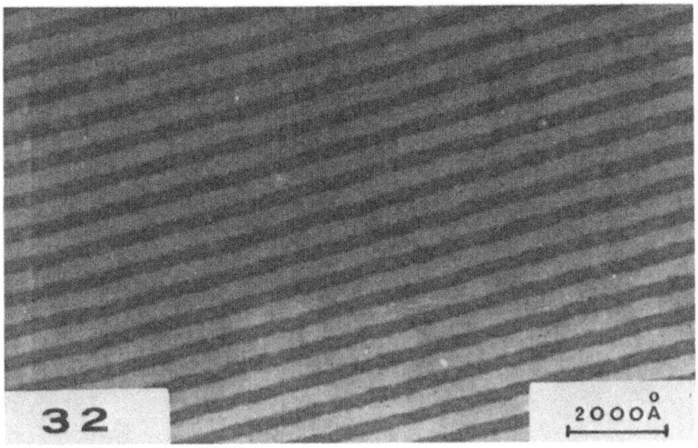

Fig. 20 (a)

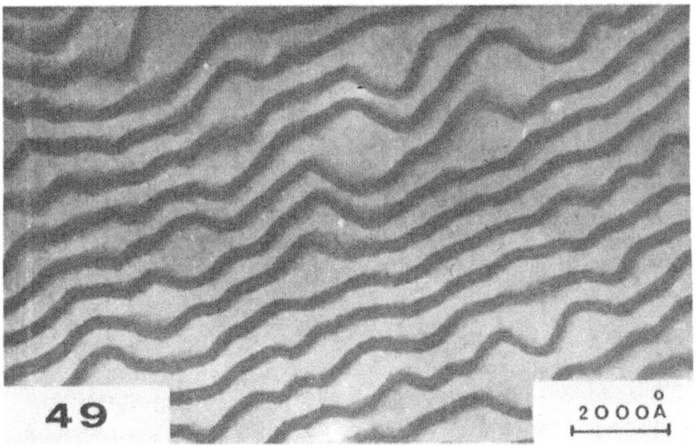

Fig. 20 (b)

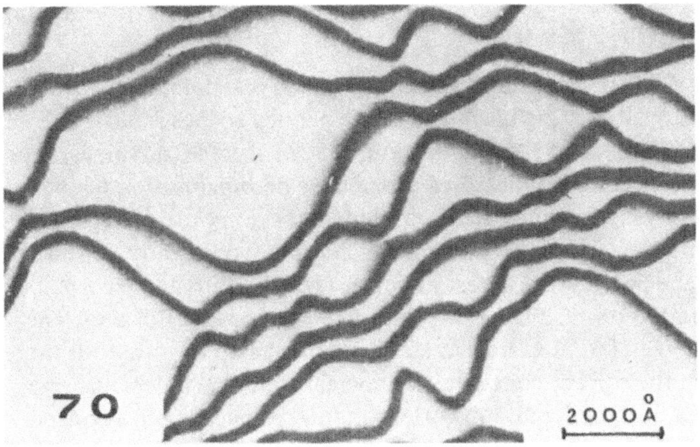

Fig. 20 (c)

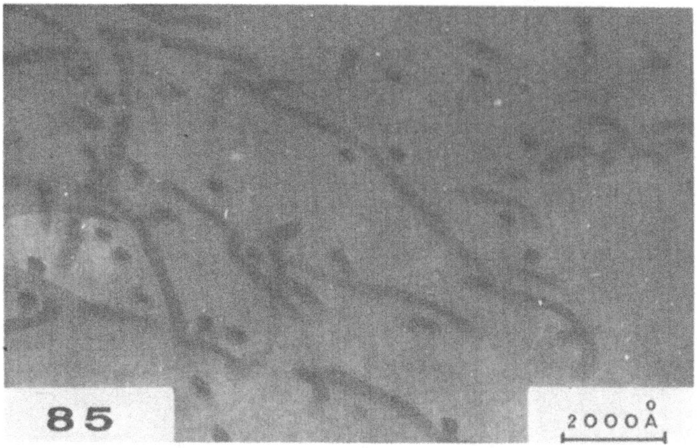

Fig. 20 (d)

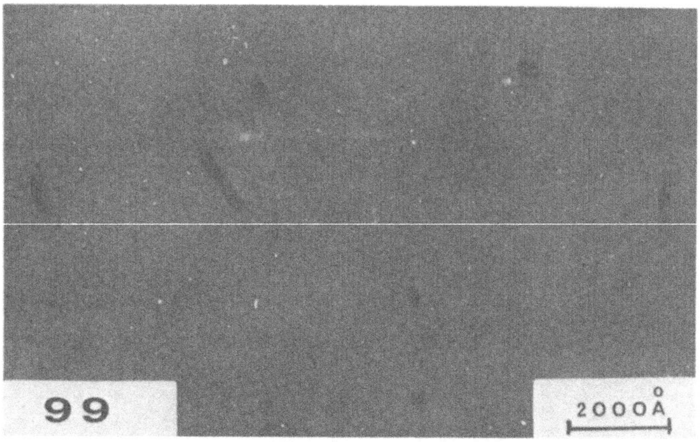

Fig. 20 (e)

Fig. 20. Desorganization of the lamellar structure by dilution. Case of the copolymer SB.11 in styrene solution. The Figs. a, b, c, d, and e correspond, respectively, to styrene concentration 32, 49, 70, 85, and 99%

Lamellar structure: For solvent concentrations between 0 and 45%, a lamellar structure characterized by alternating black and white parallel stripes is observed (see Fig. 20 a which corresponds to a concentration of 32 % styrene monomer). For solvent concentrations between 45 and 75%, a progressive desorganization of the lamellar structure is observed; the lamellae of polybutadiene do not remain planar and parallel, but the thickness of the polybutadiene layers remains nearly constant, while the thickness of the polystyrene layers increases (see Fig. 20b and c corresponding to a concentration of 49 and 70% of the added monomer). For a solvent concentration of about 75–80%, the lamellar structure breaks up to form rods more or less rectilinear. When the solvent concentration increases, the cylinders disperse and become shorter (see Figs. 20d and 20e corresponding, respectively, to concentration of 85% and 99% of the added monomer).

Hexagonal structure. For solvent concentrations smaller than about 45%, a hexagonal structure characterized by black dots (sections of polybutadiene cylinders perpendicular to the direction of their axis) arranged in a hexagonal array in a white matrix (of polystyrene) is observed (see Fig. 21a corresponding to a concentration of 35% of the added monomer). For a higher solvent concentration, a progressive dislocation of the periodic structure is observed: the size of the disordered areas increases with dilution and, for a solvent concentration of 80%, any trace of organization has disappeared and one observes only polybutadiene rods with a more or less cylindrical section dispersed at random in a polystyrene matrix; the length of the polybutadiene rods decreases with the copolymer content in the system (see Figs. 21a to 21d corresponding to monomer concentration of 35, 60, 85 and 99%).

Inverse hexagonal structure. In the inverse hexagonal structure, the solvent is located in the cylinders. For solvent concentrations smaller than about 45%, cylinders of polystyrene arranged in an hexagonal array in a polybutadiene matrix are observed (see Fig. 22a, corresponding to a concentration of 30% of the added mono-

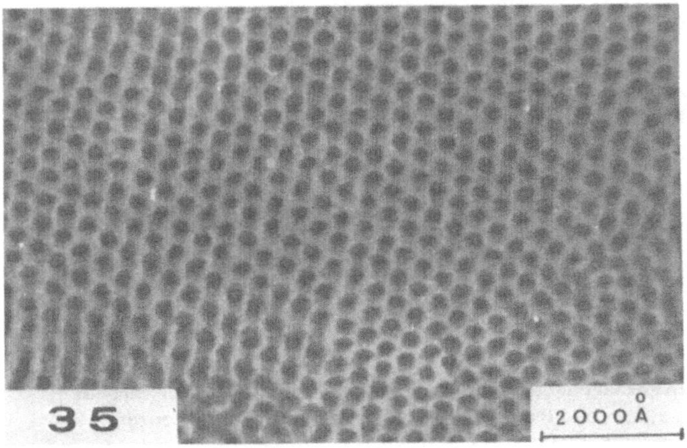

Fig. 21 (a)

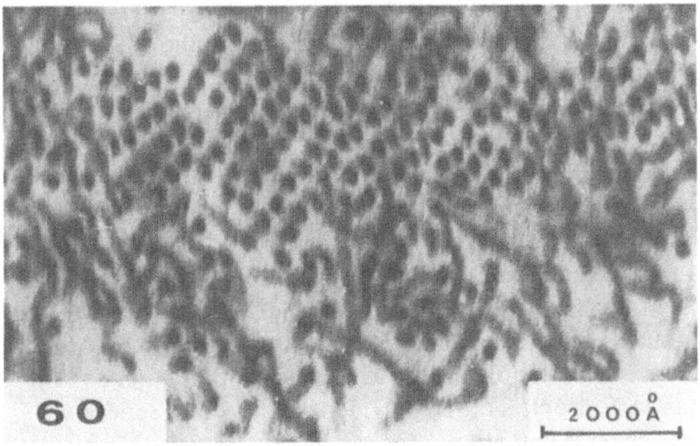

Fig. 21 (b)

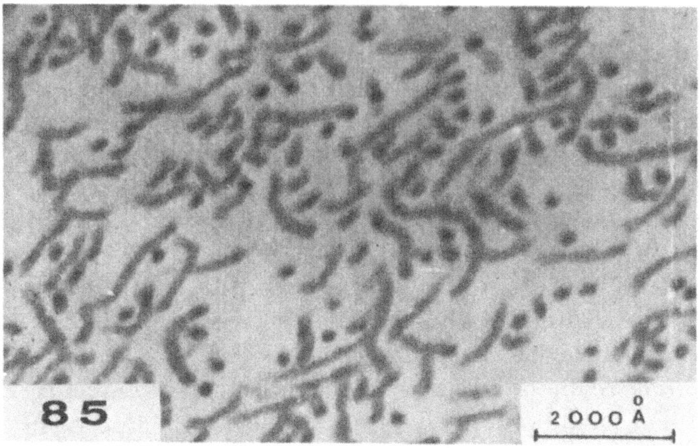

Fig. 21 (c)

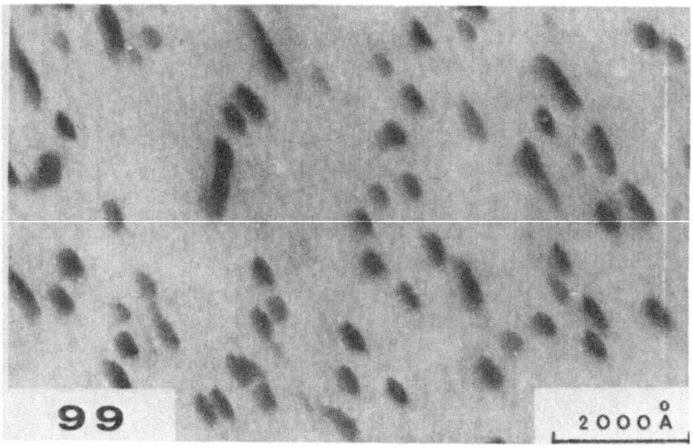

Fig. 21 (d)

Fig. 21. Desorganization of the hexagonal structure by dilution. Case of the copolymer SB. 32 in styrene solution. The Figs. a, b, c, and d correspond, respectively, to styrene concentration 35, 60, 85, and 99%

mer). For higher solvent concentrations, one observes at first a swelling of the cylinders (see Fig. 22b corresponding to 79% of the added monomer) and then a phase inversion (see Fig. 22c corresponding to 99% of the added monomer).

c 2) Case of a Monomer that is not the Monomer of one Block. The effect of the nature of the solvent on the progressive desorganization of the ordered structures of SB block copolymers has been studied in the case of the hexagonal, the lamellar, and the inverse hexagonal structure[76]. This effect is particularly striking in the case of the lamellar structure and will be illustrated by the system copolymer SB 11/methyl methacrylate.

For MMA concentrations smaller than about 45%, the structure is lamellar (see Fig. 23a, corresponding to a concentration of 28% of the added MMA). For solvent concentrations higher than 45%, droplets of the polymerized monomer appear and their size and their number increase with dilution as can be seen on Figs. 23b and

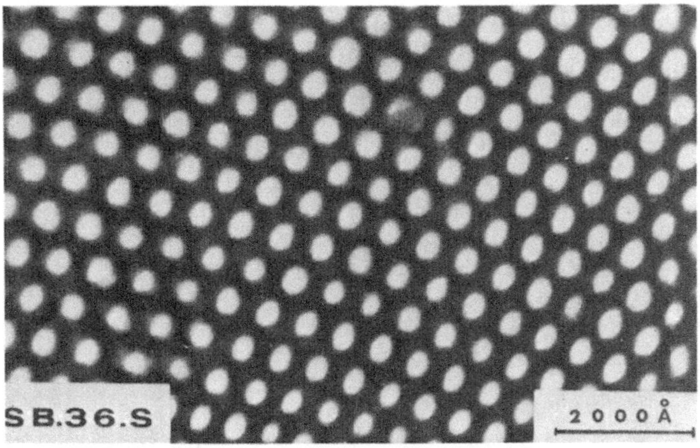

Fig. 22 (a)

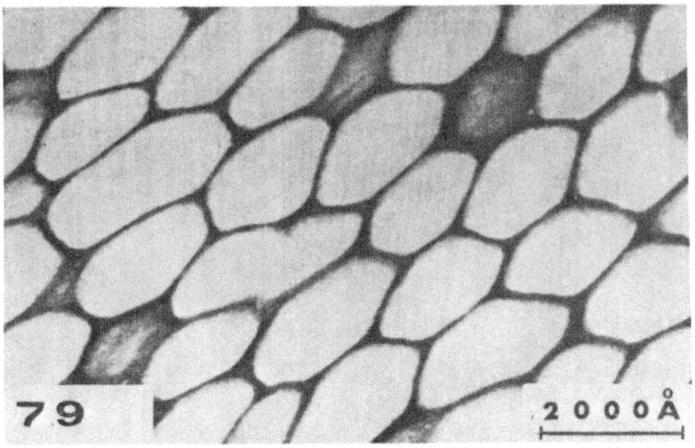

Fig. 22 (b)

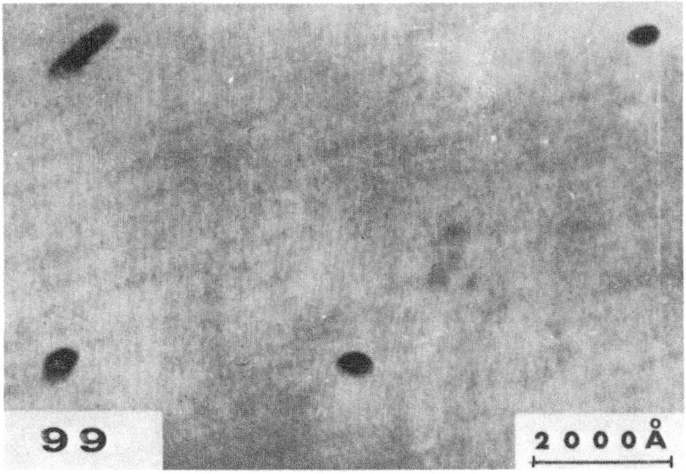

Fig. 22 (c)

Fig. 22. Desorganization of the inverse hexagonal structure by dilution. Copolymer SB. 36 in styrene solution. The Figs. a, b and c correspond, respectively, to styrene concentration 30, 79 and 99%

Fig. 23 (a)

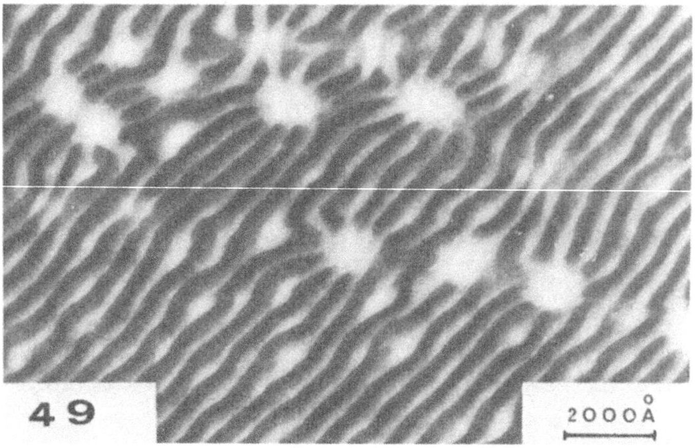

Fig. 23 (b)

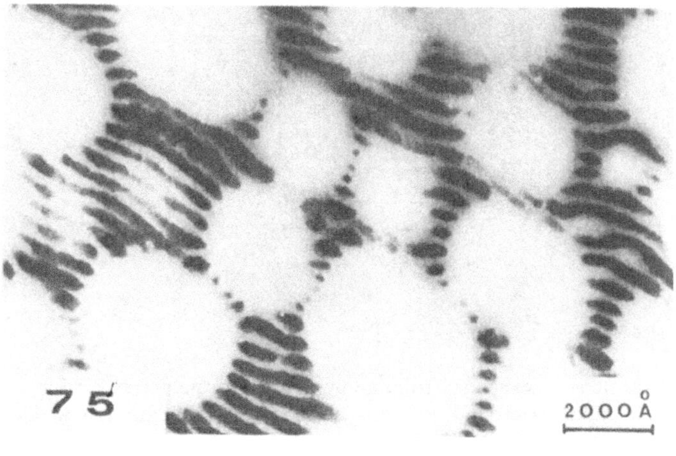

Fig. 23 (c)

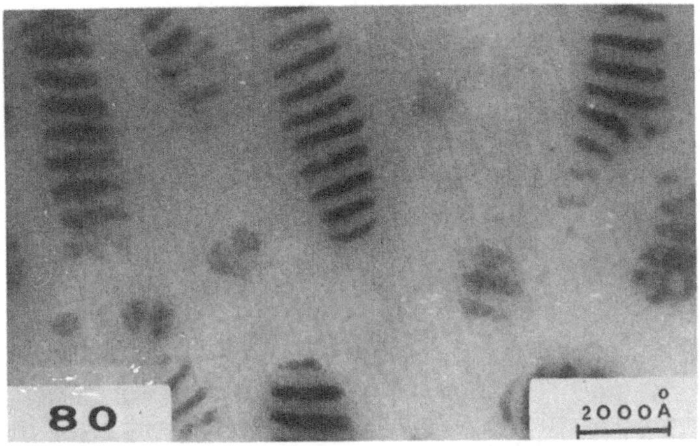

Fig. 23 (d)

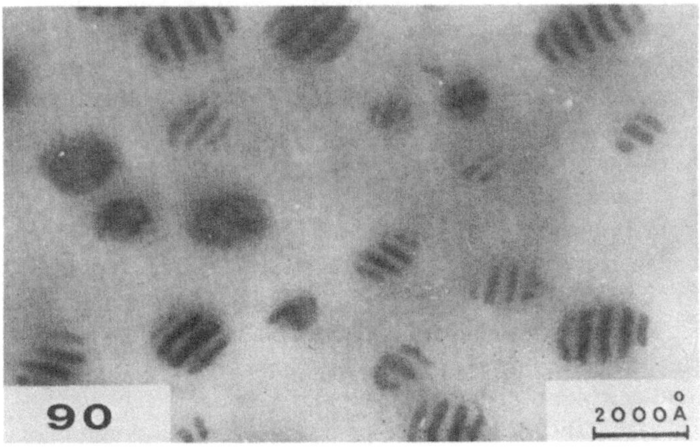

Fig. 23 (e)

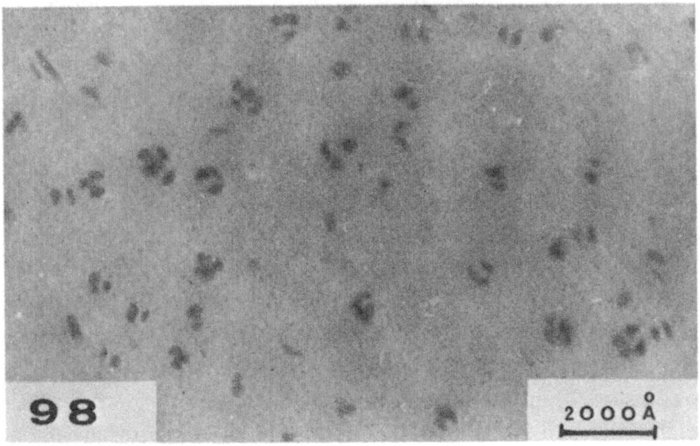

Fig. 23 (f)

Fig. 23. Desorganization of the lamellar structure by dilution. Case of the copolymer SB. 11 in MMA solution. The Figs. a, b, c, d, e, and f correspond, respectively, to MMA concentration 28, 49, 75, 80, 90 and 98%

23c, corresponding, respectively, to 49 and 75% of the added MMA. For a solvent concentration of about 80%, the polymerized MMA becomes the matrix and one observes islands formed by a rather small number of alternating layers of polybutadiene and polystyrene floating in a poly(methyl methacrylate) ocean (see Fig. 23d and 23e corresponding to 80% and 90% of the added monomer). Finally for very low concentrations of copolymer, one observes black aggregates of polybutadiene (see Fig. 23f corresponding to 99% of the added MMA).

2) Pure SB Copolymers

The most significant work on dry SB copolymers has been performed by a group of researchers of the Bayer Company[77–81]. Copolymers of various molecular weights and compositions have been synthetized by anionic polymerization in non polar solvents with butyllithium as initiator[77]; the average microstructure of the polybutadiene was 13% 1,2 and 87% 1,4 *cis* + 1,4 *trans*. The structure of these copolymers has been studied by X-rays and electron microscopy on films generally prepared by evaporation of xylene solution followed by annealing at 110 °C[78]. The German authors have tried to correlate the morphologies of the copolymers with their chemical composition. They have found spheres of polystyrene for copolymers containing less than about 15% polystyrene; a mixture of spheres and rods of polystyrene for copolymers containing between about 15% and 25% polystyrene; cylinders of polystyrene for copolymers containing between 25% and 40% polystyrene; lamellae for copolymers containing between about 40% and 60% polystyrene; cylinders of polybutadiene for copolymers containing between about 60% and 85% polystyrene; spheres of polybutadiene for copolymers containing more than about 85% polystyrene [78,79]. These results are in a good agreement with those obtained for SB copolymers with polybutadiene block of the 1,2 type in Orleans[66–76]. The Bayer researchers have also developed a theory in order to predict the type of structure of the copolymers. They assume that both types of blocks form statistical coils of similar sizes as in athermal solutions and that the supramolecular structure depends mainly on space requirements and mutual overlapping of the coils but less on the interfacial energy[62]. Using these assumptions, they have predicted the distance, forms, and sizes of the microphase domains. The predicted and experimental values of these parameters are in a good agreement[79].

Another interesting study has been performed by Vanzo who has shown that SB block copolymers of a sufficiently high molecular weight show iridescent colours (similar to those of cholesteric liquid crystals) in toluene or ethylbenzene solutions. At a concentration of 5%, the solution is colorless, at 10% it is green, at 30% it is opalescent (pink). Dried films prepared from these solutions generally retain a blue colour, and, if dried under carefully controlled conditions, the blue-green iridescent colours of more concentrated solutions may be preserved[82]. Vanzo has also studied by electron microscopy films of SB copolymers containing 60% of polystyrene and with total molecular weight between 157000 and 612000. He has shown that the structure of his copolymers was lamellar and, although the disorder of the lamellar structure was very high, he has found that the thickness of the lamellae increases with the molecular weight of the copolymer[83].

3) Block Copolymers Polybutadiene-poly(α-methylstyrene) (BMS)

Block copolymers polybutadiene-poly(α-methylstyrene) (BMS) with compositions in polybutadiene between 18 and 81% have been synthetized by anionic polymerization in tetrahydrofuran with cumylpotassium as initiator[84].

The study by low-angle X-ray diffraction and electron microscopy of concentrated solutions of the copolymers in preferential solvents for polybutadiene (isoprene, butadiene) or for poly(α-methyl styrene) (styrene, α-methylstyrene, methyl methacrylate, methylethyl ketone) and of copolymers in the dry state obtained by slow evaporation of the solvent from the mesophases have shown the existence of three types of structure: hexagonal, lamellar, and inverse hexagonal depending upon the copolymer composition[84,85]. The factors governing the structural type and the structural parameters are the same as in the case of polystyrene-polybutadiene copolymers[85].

4) Block Copolymers Polybutadiene-poly(vinyl-2-naphtalene) (BVN)

Block copolymers of butadiene and vinyl-2-naphtalene (BVN) have been synthetized and studied by the same techniques as polybutadiene-poly(α-methyl styrene) and polystyrene-polybutadiene block copolymers[86,87]. They exhibit the same structures, namely lamellar and cylindrical as SB and BMS block copolymers[86,87].

The comparison of copolymers SB, BMS and BVN with the same compositions and molecular weight and exhibiting all a lamellar structure has shown that the extension of the macromolecular chains depends upon the nature of the nonpolybutadiene blocks. The extension of the chains increases in the order styrene, α-methylstyrene, vinylnaphtalene and the extension of the polybutadiene chain is caused by that of the adjoined chain[85].

5) Block Copolymers Polybutadiene-polystyrene-polybutadiene (BSB)

Symmetric BSB copolymers containing between 20 and 80% polybutadiene and covering a large range of molecular weights have been synthetized by Douy in tetrahydrofuran solution at a low temperature (−70°) using a bifunctional initiator (α-methylstyrenepotassium dianion)[66,88].

The study by low-angle X-ray scattering, electron microscopy, and differential scanning calorimetry of the mesophases obtained by dissolution of BSB copolymers in preferential solvents for the polystyrene block and of dry BSB copolymers obtained by slow evaporation of the solvent from the mesophases has allowed to establish the respective effect of different factors which control the structure of the mesophases and their geometrical parameters. It has been shown that the nature, concentration, polymerization of the solvent, and temperature have the same effect on BSB copolymers as on SB copolymers[35,88–91].

The study of sets of copolymers with a type of block (polystyrene, or polybutadiene) of constant molecular weight and the other of variable molecular weight has shown that, within the stability region for a given structure, the conformation of the insoluble polybutadiene blocks is independent of the length of the soluble polystyrene blocks, while the conformation of the soluble polystyrene chains depends upon the polybutadiene content in the copolymer. The laws of variation of the diameter of the polybutadiene cylinders in the case of the hexagonal structure and for the polybutadiene layer thickness in the case of the lamellar structure have been established[88] as a function of the molecular weight of the polybutadiene blocks.

The comparison of BS and symmetric BSB copolymers derived from the BS polymer by addition of a B block identical to the first one has demonstrated that the addition of a third block to a diblock copolymer achieves a stretching of both the polybutadiene and the polystyrene chains[92].

6) Block Copolymers Polystyrene-polybutadiene-polystyrene (SBS)

Symmetric SBS block copolymers covering a wide range of compositions and molecular weights have been synthetized[66] and studied by the same techniques as symmetric BSB copolymers[93–95]. In solution in methylethyl ketone, methyl methacrylate, vinyl acetate, or styrene they exhibit a behaviour similar to that of SB and BSB copolymers[93–95] with respect to the effect of temperature, concentration, and postpolymerization of the solvent. The effect of the molecular weight of the soluble and insoluble blocks on the geometrical parameters of the hexagonal and lamellar structures is however different for BSB and SBS copolymers. For SBS copolymers, there is a reciprocal interaction between soluble and insoluble blocks. For instance, the thickness of the soluble and insoluble layers in the lamellar structure depends upon the molecular weight of both the soluble and insoluble blocks[95].

7) Pure SBS Copolymers

In 1965, thermoplastic elastomers were introduced by Shell Chemical Co and marketed under the trademark Kraton. Thermoplastic elastomers are ABA block copolymers, where A is polystyrene and B is polybutadiene or polyisoprene. Kratons exhibit at service temperatures mechanical properties similar to those of conventional reinforced rubber vulcanizates such as snappy return form high elongation, high resilience, good tensile strength, and low set. At higher temperatures, they undergo a transition to a melt and behave as thermoplastic materials. The elimination of the vulcanization step allows the use of high speed injection moulding and extrusion and the reuse of scrap[7–10]. These original technological properties have induced a lot of fundamental studies of ABA copolymers in order to determine relations between the chemical structure, morphology and mechanical properties of the materials.

a) First Studies

Among the first studies of SBS copolymers we have to mention the work performed by Hendus *et al.*[96]. They studied sets of copolymers where block ratios and molecular weights were varied systematically. Electron micrographs of cast films and sections of compression moulded samples have shown that the minority block component always formed the dispersed phase and the other the matrix.

Aggarwal[97–99] has been a pioneer in the study of relations between morphology and properties of ABA copolymers. He studied at first films of Kraton 101 cast from 90/10-THF/MEK solutions, carbon tetrachloride solutions, and 90/10-benzene/heptane solutions by different technics[97]. The vibrating-reed apparatus revealed, in addition to the glass transition, an intermediate broad damping peak near 30 °C for samples cast from CCl_4. This intermediate damping peak may be due to a glass transition of polystyrene segments that are not completely separated from the polybutadiene phase and would indicate an appreciable amount of phase mixing in samples cast from CCl_4. The samples prepared from a benzene/heptane solution gave evidence of a broad intermediate transition. The samples cast from a THF/MEK solution showed the largest and best defined polystyrene transition and only a small intermediate transition. The stress-strain curves confirmed the influence of the nature of the solvents used for casting films. The stress-strain curves of specimens prepared by solution casting from THF/MEK solvent systems exhibited a yield point at about 3% elongation. Stress-strain curves for specimens cast from benzene/heptane solutions were more "rubber-like". Stress-strain curves for specimens cast from CCl_4 solutions were more "leathery" than those of films cast from other solvents. Electron micrographs revealed at first that polystyrene chains form aggregates of about 120 Å in diameter for all systems studied, but they also showed that the nature of the casting solvent produces changes in morphology which are subtle but reproducible. Electron microscopy allowed at last the description of the deformation process of the glassy polystyrene network. The high values of the Young modulus and the stress-softening effects of the stress-strain curves were explained assuming a polystyrene continuous phase whose weaker ties are progressively broken by increasing load.

b) Bristol Group's Studies

Probably the most significant studies on the morphology of pure SBS copolymers were performed by Keller and collaborators[100–112]. As their results have been recently reviewed by Keller himself[36], we shall not give a detailed analysis of their research but only sum up their contribution.

Keller *et al.* have studied plugs of Kraton 1102 (the molecular characteristics of this SBS copolymer containing 26% by weight of polystyrene are listed in Table 4) obtained by extrusion under vacuum at temperatures 100–120 °C and under a constant pressure of approximately 8.5 $kgcm^{-2}$ [100] followed by annealing. Low-angle X-ray patterns obtained with the beam oriented parallel and perpendicular to the extrusion direction have revealed the existence of a macro "single crystal" formed by an hexagonal lattice of long polystyrene cylinders parallel to the extrusion direction. The hexagonal lattice periodicity is 300 Å and the cylinder diameters 150 Å (cf. [101–102]). Electron micrographies of sections perpendicular and parallel to the extrusion direction have confirmed the X-ray results: sections perpendicular to the extrusion direction are characterized by white spots (sections of polystyrene

Table 4. Molecular characteristics of Kratons 1101 and 1102 according to Refs.[113] and [114]

	K 1101	K 1102
Total molecular weight $\overline{M}_w$	102×10^3	75×10^3
% Polystyrene	33	26
Terminal PS block M_w	17×10^3	10×10^3
Central PB block M_w	68×10^3	55×10^3
Microstructure:		
PS 1,4 *trans* %	42	48
PS 1,4 *cis* %	49	44
PS 1,2%	9	8

cylinders) in a black matrix while sections parallel to the extrusion direction are characterized by alternating dark and light parallel stripes[103].

The study of the birefringence of the Kraton "single crystal" and of the effect of stress applied along and perpendicular to the extrusion direction has shown that the birefringence is entirely due to form birefringence and that the dispersed styrene phase and the butadiene matrix consist of randomly oriented chains[104]. This results have been confirmed by infra-red dichroism studies[105].

The study of the Young modulus carried out on samples cut with their long axis at different orientations with respect to the extrusion direction has shown that the single crystal behaves as a glass along the cylinder direction and as a rubber in the perpendicular direction[104]. The comparison of the dependence of Young modulus on the angle between the stress direction and the original extrusion direction are in agreement with the predictions of current theories of fibre reinforcement based on hexagonal symmetry if the Poisson ratio is taken 0.37 instead of the bulk value 0.50 (cf. [106]).

Later, Dlugosz, Folkes, and Keller have studied by low-angle X-ray diffraction, electron microscopy, and electron diffraction a SBS copolymer (TR 41 1649) richer in styrene (48.2% by weight) oriented by their extrusion method. They have found a lamellar structure with intersheet spacings of 260 Å and a polystyrene layer thickness of 120 Å [107]. As far as it can be judged from the X-ray patterns and electron micrographs the order is much better in the extruded sample than in both the unannealed and annealed raw material, but it is by far not so good as in extruded Kraton 1102.

Recently[108], Keller *et al.* have examined the optical and swelling properties of the SBS copolymer TR 41 1649 from Shell Company (polystyrene content 48.2% by weight; structure lamellar) oriented by a modification of their extrusion method called "radial flow method". They have studied the birefringence and the dimensional changes of their samples as a function of the volume fraction of swelling agent. The swelling agents used – octane, decane, dodecane – were solvents of the polybutadiene blocks. Keller *et al.* have found that the observed birefringence is principally due to the form birefringence, but a small contribution from a slight molecular orientation of the polystyrene chains perpendicular to the lamellar surfaces has been revealed by infrared dichroism.

Similar studies on dimensional changes and birefringence have been performed on Kraton 1102 (polystyrene content 26% by weight; structure: hexagonal lattice of polystyrene cylinders parallel to the extrusion directions)[109]. The swelling agents used were all solvents for the polybutadiene matrix (hexane, octane, decane). The swelling behaviour may be divided in two regions: the reversible one, where the sample recovers its original dimensions by drying, and the irreversible one, where after drying the longitudinal dimension (parallel to the cylinder direction) remains higher than in the unswollen state and the lateral dimension decreases to a value smaller than its initial unswollen value. In the reversible region, the swelling is in agreement with a lateral expansion of the hexagonal lattice of cylinders. In the irreversible region, both low-angle X-ray results and electron microscopy[110–111] results are in favour of a modification of the original structure.

Finally Keller[112] has developed a theory to explain the deformation behaviour of SBS copolymers with an hexagonal structure formed by polystyrene cylinders in a polybutadiene matrix. His theory, called "Random-break treatment" is based on a model consisting of a system of broken cylinders of the same initial length with the breaks distributed randomly along the cylinders. The author regards the cylinders unstrained compared with the matrix material. He assumes the matrix to consist of small Hookean elements joining each rod to the six nearest rods and he calculates the stress in the cylinders due to the tensile strain of the matrix. The theoretical predictions are in a good agreement with the birefringence and electron microscopy results.

c) Genoa Group's Studies

Pedemonte *et al.*[113] have also obtained "a macrolattice" with Kraton 1101 (the molecular characteristics are given in Table 4). Kraton 1101 was extruded at 0.5 mm/mn at 180–200 °C and 8 kg/cm^2 [113] or at 220 °C and 9.7 kg/cm^2 [34] and annealed at 150–170 °C [34, 113]. Sections parallel and perpendicular to the extrusion direction have been obtained by ultramicrotomy at low temperature and electron micrographs have given the following morphologies. The original sample is formed by polystyrene cylinders of various lengths and orientations. The annealed sample is polycrystalline and presents small domains of polystyrene cylinders arranged in an hexagonal array. The extruded samples reveal a beginning of orientation of the polystyrene cylinders parallel to the extrusion direction. The extruded and annealed sample forms a "macrocrystal" of polystyrene cylinders with their axis parallel to the extrusion direction and arranged in a hexagonal array[113]. The cylinders are larger and the number of dislocations is smaller than in the case of Kraton 1102, but it is more difficult to obtain a macrolattice with K. 1101 than with K. 1102 the molecular weight and the length of the polystyrene blocks being higher and the mobility of the macromolecular chains being smaller for K. 1101.

The Italian authors have also studied the effect of the annealing and of the solvent evaporation rate on the morphology and on the mechanical properties of different Kratons[114–117]. Kraton 1102 exhibits a Mullins effect which increases with annealing. Films with thickness between 0.2 and 0.4 mm, cast from dilute solution (1%) in cyclohexane and methylethyl ketone (MEK) at different rates have been studied[117]. Electron micrographs have revealed that the morphology of films cast from cyclohexane solution is nearly independent of the evaporation rate while the morphology of films cast from MEK depends upon the evaporation rate. The stress-strain curves obtained with all films exhibit a Mullins effect which is higher for film cast from MEK. The low-angle X-ray patterns show a structural deformation during elongation.

The stress-softening effect exhibited by SBS copolymers has been correlated with their morphology[118]. Samples of K. 1102 were extruded and annealed for 10 days at 150–170° then they were swollen for 30 minutes or 2 hours in n-heptane and dried under vacuum at room temperature until a constant weight was obtained. Before and after swelling, stress-strain curves were dawn and longitudinal sections were cut. Stress-strain results have shown that swelling in n-heptane decreases both the Young modulus and the yield stress: the stress softening has completely disappeared after a swelling for two hours but has only decreased after a swelling for 30 minutes. Furthermore, the swelling reduces the stress required to reach any deformation level and annealing of previously swollen samples has no effect. Electron micrographs obtained before swelling of the sample are characteristic of a macrolattice of polystyrene cylinders parallel to the extrusion direction and hexagonally packed. After swelling of the sample, electron micrographs reveal a polycrystalline material where small well ordered domains are randomly oriented. In the original extruded and annealed material, the polystyrene cylinders were practically continuous with only some dislocations. The stretching causes first the formation of a large number of thin ties along the axis of the polystyrene rods and the cylinders become similar to strings of pearls. Subsequently, the break of the cylinders occurs. Therefore, the stress-softening effect results from a disruption of the polystyrene continuous phase according to the hypothesis of Aggarwal[97].

Pedemonte *et al.* have performed a detailed study of the dependence of the preparation of samples on their morphology and stress-properties[34, 119]. For K. 1101 they have compared the original copolymer, films cast from toluene solution at two different evaporation rates (about 20 and 0.5 cm^3/h), compression moulded films, and extruded and extruded-annealed specimens. From annealing studies, it has been concluded that the original material contains rod-like polystyrene domains. From the comparison of the electron micrographs and the stress-strain curves of both extruded and extruded-annealed samples, the following conclusions have been drawn: the high values of the Young modulus are caused by the continuous polystyrene phase resulting from the arrangement of the polystyrene rods along the extrusion axis; the higher stiffness of the extruded-annealed plugs is due to the improvement of the phase separation resulting in a more regular and continuous distribution of the polystyrene cylinders; the yield point is explained by the presence of many dislocations and thin ties which link consecutive cylinders; the stress-softening is another consequence of the tie breaks whose number increases with the load. In the case of solution cast films, the morphology of samples prepared at a high evaporation rate does not show any regular arrangement of the polystyrene domains which seem to have a rod-like shape, while for low rates, a morphology similar to that of the original annealed samples is observed[113]; first deformation curves of fast and slowly evaporated films are not too different and do not show a well defined yield point, while second deformation curves exactly coincide. In moulded films, the polystyrene chains form rod-like domains in a rubbery matrix but no particular orientation of the cylinders exists; the cylinders are oriented by stretching.

In order to correlate the morphology of SBS copolymers with the hardening effect observed at the highest extension ratios after a rapidly repeated deformation, Pedemonte *et al.* [118, 120] have studied the stress-strain and birefringence curves of Kraton 1107 (a SIS copolymer containing 10% of polystyrene), Kraton 1101 and Kraton 1102. Samples of Kraton 1107 were obtained by a slow evaporation of cyclohexane from cast films or by compression moulding; the samples exhibit an irregular distribution of well defined polystyrene spheres in a polyisoprene matrix. Samples of Kraton 1101 were cast films obtained by a slow evaporation and compression moulded films; they exhibit a polycrystalline structure in which polystyrene cylinders are hexagonally packed in a rubbery matrix. Samples of K 1102 were extruded filaments, they consist of polystyrene cylinders hexagonally packed and oriented along the extrusion axis. It has been found that the hardening effect is independent of the chemical nature of the elastic central block (polybutadiene or polyisoprene), amount of polystyrene (at least between 10 and 33% by weight), molecular weight and length of the blocks, technique used for the specimen preparation (extrusion, compression moulding, solvent casting), and morphology of the specimen. On the contrary, the hardening effect is a function of the time ellapsed between two successive deformations and the hardening effect is due to a kinetic effect. A model has been proposed which involves the recovery of the original molecular conformations of the elastic central block between two successive deformations and the hardening has been explained assuming that elastic chains in the specimen are partly stretched.

d) Other Studies

A number of other studies have been performed on Kratons and similar SBS copolymers. Lewis and Price have studied the morphology of Kraton 1101 and 1102[33,121,123]. Samples prepared by compression-moulding at 152 °C were studied by low-angle X-ray diffraction[122]. Films cast from dilute benzene solution on a mercury surface at different rates of evaporation were studied by both X-ray diffraction and electron microscopy[33,122]. A hexagonal lattice of polystyrene cylinders embedded in a polybutadiene matrix was observed for compression moulded samples and for films cast at low rates. On the contrary, an irregular texture was observed for films prepared under flash conditions; decreasing rates had provided a greater domain ordering. Force extension curves have revealed an anisotropy of mechanical properties for compression-moulded samples but an isotropy for cast films[122].

Fisher[124,125] has studied the stress-strain and optical properties of three SBS block copolymers containing, respectively, 31, 40 and 49% polystyrene as a function of temperature. He has shown that these materials are two phase systems in which the polybutadiene chains form an elastomeric phase and the polystyrene chains a glassy phase acting as physical crosslinks. Fisher[126] has also obtained electron micrographs of toluene cast films showing a hexagonal arrangement of polystyrene domains and its deformation under a stretching of 50%.

Matsuo *et al.* [127,128] have studied dynamic mechanical properties (dynamic modulus and dynamic losses), stress-strain behaviour, elongation at break, stress relaxation, impact strength, light transmittance, heat distorsion temperature, and flow properties of a set of SBS copolymers containing, respectively, 80, 70, 60, 50, and 40% polystyrene and SB, BSB, and BSBS copolymers containing all 60% polystyrene in order to show the effect of the composition and the geometry of the copolymer on its technological properties. Matsuo *et al.* have also studied by electron microscopy ultrathin sections of compression moulded sheets; they have observed a phase separation but no ordered structures. Later Matsuo *et al.* [128] have studied sections cut parallel and perpendicular to the surface of films and obtained by evaporation of an approximately one per cent toluene solution. Although the structure presented by their micrographs were very disordered, they had been able to find a spherical, cylindrical, or lamellar morphology and to show changes of the morphology as a function of the copolymer composition.

Kraus has compared the dynamic properties (storage moduli, loss tangents) and the glassy transition temperatures of random and block copolymers and related them to the block length and the compositional purity of the blocks [130]. He has related the viscoelastic properties and tension shear of films cast from good (toluene, CCl_4) and poor solvents for polybutadiene (ethyl acetate, methylethyl ketone) and of compression-molded films with the morphology of SBS copolymers (containing 30% by weight of polystyrene) revealed by electron microscopy [131]. For copolymers SBS, BSB, and SIS containing about 50% polystyrene by weight but of different molecular weight, he has related the dynamic viscoelastic behaviour (dynamic storage modulus and dynamic loss moduli) to the dimensions and the composition of the domain boundary; he has interpreted his results by decreasing the existence of a diffuse interphase whose dimensions increase with decreasing block lengths [131].

The Thermoelastic 125 (30% by weight of polystyrene) has been studied by Canter [132] as a function of temperature using a specimen moulded at 150 °C. Differential scanning calorimetry has revealed the presence of two transitions. The lower transition temperature corresponds to T_g of polybutadiene chains and the upper transition temperature corresponds to T_g of styrene chains. The upper transition has also been studied by temperature dependence of the torsional modulus.

The dynamic viscoelasticity and the thermal behaviour of films of Thermoelastic 125 cast from solutions in four solvents – toluene (T), carbon tetrachloride (C), ethyl acetate (E), and methyl ethyl ketone (M) – have been studied by Miyamato [133]. The mechanical loss tangent (tan δ) and the storage modulus E' dependences exhibit two transitions at –70 °C and 100 °C which have been attributed to onset of motion of polybutadiene and polystyrene segments, respectively. The heights of the polybutadiene peaks on tan δ curves decrease in the order $C > T > E > M$, while for polystyrene the order is reversed: $C < T < E < M$. These phenomena have been related to the magnitude of phase separation of the polystyrene and polybutadiene blocks.

The viscoelastic and ultimate tensile properties of Kraton 101 (SBS copolymer containing 30.5% polystyrene and with total molecular weight 76000) and of Thermoelastic 226 (a similar material which contains about 35% of plasticizer and inorganic pigments) have been studied by Smith[135] using samples moulded in a hydraulic press at 140–150 °C. The properties of the Kraton 101 and the Thermoelastic 226 have been compared with those of SBR vulcanizates containing 0.15 and 25% by volume of polystyrene spheres. It has been deduced[135] that the time dependence of their mechanical properties and their high tensile strength result from energy dissipation associated with the plastic deformation and eventual disruption of the colloidal polystyrene domains.

The mechanical and rheo-optical properties of Kraton 101 have been studied by Stein[136] using films cast from methylethyl ketone and from toluene solutions. The stress-strain curves, birefringence-strain curve, stress relaxation, birefringence relaxation, and dynamic mechanical spectra are dependent upon the morphology of the copolymer which in turn is dependent upon the conditions of preparations of the samples.

Kaelble has developed a model[137] to relate mechanical properties of SBS and SIS copolymers to their interfacial morphology. The adsorption-interdiffusion model for the interfacial phase defines the size, shape, and connectivity of microdomains. Kaelble has applied his model to the interfacial morphology in order to explain the initial tensile yielding, cold drawing, and subsequent hysteresis in recovery of Kraton 101[138,139].

Using Porod's two-phase theory of small-angle X-ray scattering Kim[140] has analyzed the intensity data for an unoriented molded sample of Kraton 101 and found that the interface between the styrene and butadiene phases is sharp.

Hourston[142] has studied the effect of casting solvents on some physical properties of two SBS copolymers which seem to be Kraton 1101 and 1102. The properties of films cast from cyclohexane solution were found to be independent of the evaporation rate while those of films cast from toluene solution were found to be modified by the evaporation rate.

Interesting studies of the structure and properties of SBS copolymers have been performed in the Morton's group[143–146], but the sturctural part of their work has been seriously questioned by Skoulios[158] and Pedemonte[119].

Many other studies have been devoted to SBS copolymers and some are reported in references[147–152].

8) Star Copolymers of Styrene and Butadiene

To our knowledge, very few results have been published concerning the morphology of star SB copolymers. In 1975, Pedemonte *et al.*[153] studied two copolymers $(SB)_4Si$ called Europrene T 161 and Europrene T 162 containing, respectively, 35% and 49% polystyrene. For the annealed samples, a disordered cylindrical structure was found in T 161 and a lamellar structure in T 162. In 1976, Pedemonte *et al.*[154] published the results of a more detailed study performed on compression-moulded films of T 162. These films deformed increasing step by step the maximum strain value exhibit the Mullins effect. If in the original compression moulded film rods of polystyrene were arranged almost perpendicular to the compression plane, the deformation of the specimen produced a reorientation of polystyrene rods: the polystyrene rods were oriented in the stretching direction, which was parallel to the compression plane.

Kraus[155] has studied the steady flow and dynamic viscosity of the following branched butadiene-styrene block copolymers $(BS)_3$, $(SB)_3$, $(SB)_4$ in comparison with BSB and SBS copolymers. He has found higher viscosities (at constant molecular weight and total styrene content for polymers terminated by styrene blocks) for the former irrespective of branching, but for copolymers of equal molecular weight the viscosity is smaller for branched than for linear copolymer. Kraus[156] has also studied the effect of free polybutadiene molecules on the viscoelastic behaviour of branched $(SB)_4$ block copolymers which consist of styrene domains in a butadiene matrix and verified De Gennes's theory of "reptation"[157].

B) Block Copolymers of Isoprene and Other Monomers

Among copolymers containing isoprene, AB and ABA block copolymers of styrene and isoprene have been the most studied, but interesting results have also been obtained with copolymers of isoprene and vinylpyridine and of isoprene and methyl methacrylate.

1) Mesophases of AB and ABA Block Copolymers of Styrene and Isoprene

The study by low-angle X-ray diffraction of AB and ABA block copolymers of styrene and isoprene has allowed Gallot *et al.*[159–166] to demonstrate that the existence of mesomorphic structures is a quite general property of block copolymers; this property is related to the incompatibility of the macromolecular chains[11] and the presence in the molecule of a crystallizable (*e.g.* poly(ethylene oxide)) block is not the condition that governs the formation of mesomorphic structures.

Copolymers SI and SIS of a wide range of molecular weights and compositions have been synthetized by anionic polymerization, under vacuum, at room temperature, in toluene solution, using sec.-butyllithium as initiator[167].

a) Structures Observed

In toluene or styrene solution, SI and SIS copolymers exhibit a behaviour characteristic of copolymers swollen by a non-selective solvent[11,35]. In fact, the partition coefficient of the solvent is slightly in favour of polystyrene ($\varphi_A = \varphi_{PS} = 0.6$ as follows from the respective intensities of the different diffraction orders in low-angle X-ray patterns)[75]. Depending upon the copolymer composition, lamellar and hexagonal structures have been observed. In the case of SI and SIS copolymers in toluene or styrene solution, it is not necessary to distinguish between the hexagonal and the inverse hexagonal structures, the two types of blocks being nearly equally swollen[11,35]. When the solvent concentration increases, all parameters of the lamellar structure of SI copolymers increase. On Fig. 24, one can see that with increasing toluene concen-

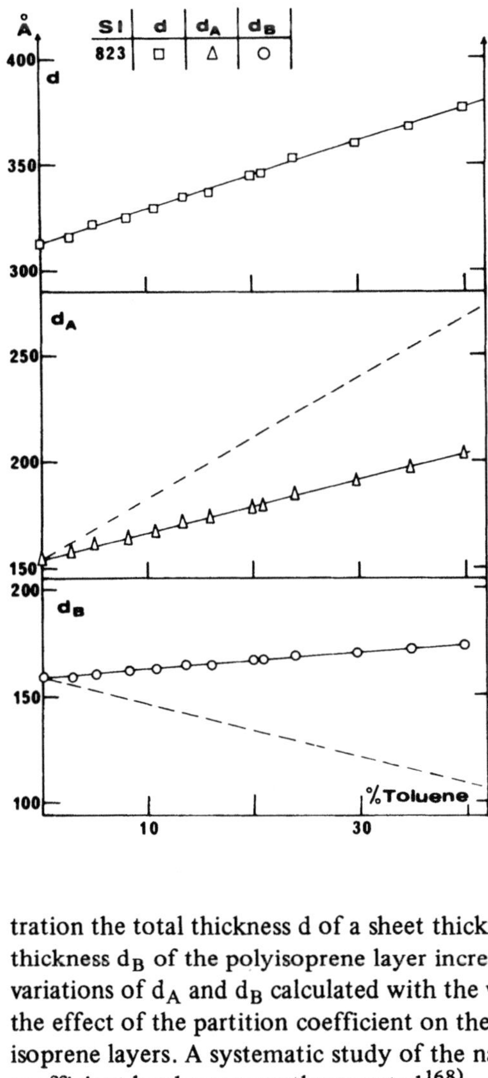

Fig. 24. Variation of the geometrical parameters of the lamellar structure of the copolymer SI. 823 containing 52% polystyrene with toluene concentration. Toluene is a non-selective solvent, $\varphi_A = 0.6$.
□: d = total thickness of a sheet; △: d_A = thickness of the polystyrene layer;
○: d_B = thickness of the polyisoprene layer; doted lines: d_A and d_B calculated with the hypothesis $\varphi_A = 1$

tration the total thickness d of a sheet thickness d_A of the polystyrene layer, and thickness d_B of the polyisoprene layer increase. The dotted lines in Fig. 24 show the variations of d_A and d_B calculated with the wrong hypothesis $\varphi_A = \varphi_{PS} = 1$ to explain the effect of the partition coefficient on the thickness of the polystyrene and polyisoprene layers. A systematic study of the nature of the solvent on the partition coefficient has been recently reported[168].

b) Influence of Molecular Weight and Copolymer Composition

For SI copolymers, the total thickness ($d = d_A + d_B$) of the lamellar structure increases with the total molecular weight of the copolymer (Fig. 25). Furthermore, the study of a set of copolymers with a constant molecular weight of the polystyrene block (23000) and a variable molecular weight for the polyisoprene block (from 8000 to 28000) has shown that the thickness d_B of the polyisoprene layer increases with the molecular weight of the polyisoprene block M_B (Fig. 26), while the thickness d_A of the polystyrene layer is independent of the molecular weight of polyisoprene (Fig. 27). The study of a set of copolymers with the same polyisoprene block (17500) but a variable polystyrene one has shown that the thickness d_A of the

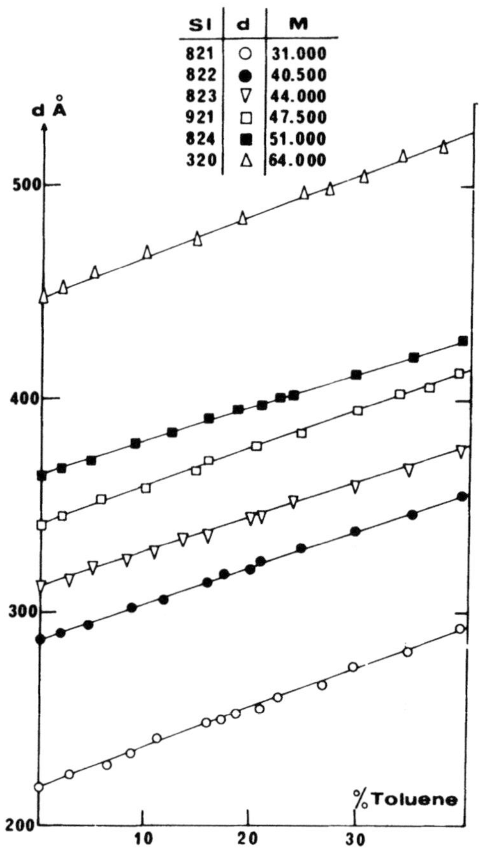

Fig. 25. Influence of the molecular weight of the copolymer on the total thickness d of a sheet of the lamellar structure.
○: SI. 821 and M = 31 000; ●: SI. 822 and M = 40 500; ▽: SI. 823 and M = 44 000; □: SI. 921 and M = 47 500; ■: SI. 824 and M = 51 000; △: SI. 320 and M = 64 000

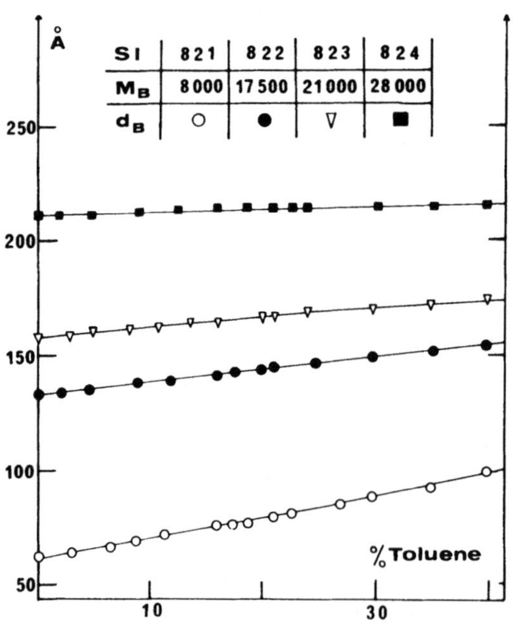

Fig. 26. Influence of the molecular weight M_B of the polyisoprene block on the thickness d_B of the polyisoprene layer.
○: SI. 821 and M_B = 8000; ●: SI. 822 and M_B = 17 500; ▽: SI. 823 and M_B = 21 000; ■: SI. 824 and M_B = 28 000

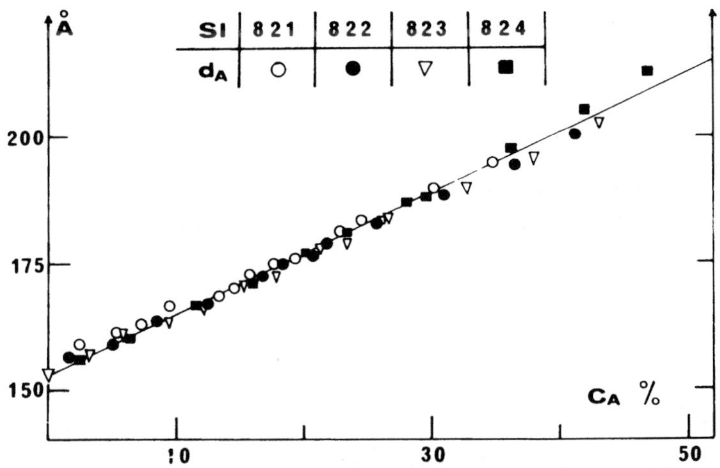

Fig. 27. Influence of the molecular weight M_B of the polyisoprene block on the thickness d_A of the polystyrene layer.
C_A: swelling ratio of the polystyrene block

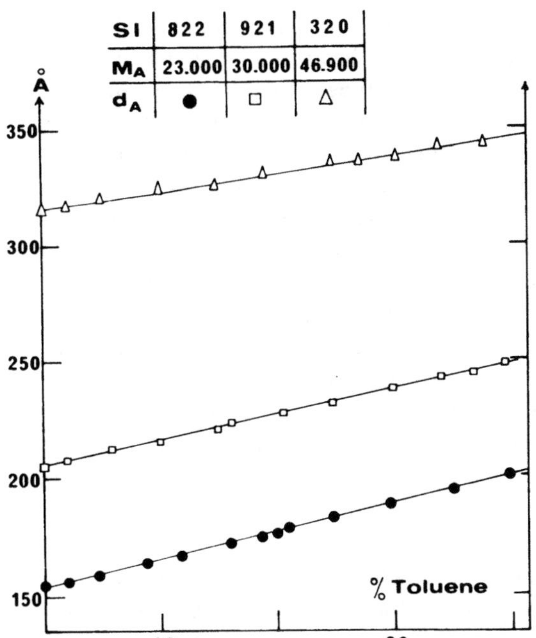

Fig. 28. Influence of the molecular weight M_A of the polystyrene block on the thickness d_A of the polystyrene layer. •: SI. 822 and M_A = 23000; □:SI. 921 and M_A = 30000; △: SI. 320 and M_A = 46900

polystyrene layer increases with the molecular weight of the polystyrene block (Fig. 28).

c) Effect of the Number of Blocks

SIS and SI copolymers behave similarly and the addition of a second S block to an SI copolymer has the same effect on the chain conformation like the addition of a polybutadiene block to a BS copolymer (cf.[92]).

2) Pure SI and SIS Copolymers

In the case of block copolymers from styrene and isoprene, it is not necessary to review separately the results obtained with bi an triblock copolymers for the following reasons: first, the relations between the morphology and the mechanical properties of SIS copolymers have not been studied as extensively as those of SBS copolymers and the results obtained are similar, second, morphological studies have been performed by the same authors on both SI and SIS copolymers.

Kawai and collaborators have extensively studied copolymers of styrene and isoprene. They have synthetized and studied SI copolymers[63,169], SIS copolymers[64], and ternary polymer blends of SI block copolymers with polystyrene and polyisoprene[170–171] and proposed a thermodynamic interpretation of the domain structure in solvent-cast films of SI, SIS, and blends of these copolymers with their homopolymers[63,64,172].

Kawai *et al.* have studied films of SI copolymers of different composition obtained by evaporation of about 5% toluene solution. Electron micrographs of sections perpendicular to the film surface have revealed four types of morphology: spheres of polyisoprene randomly distributed in a polystyrene matrix for a polystyrene content of 73 wt.-%, a rather disordered lamellar structure for polystyrene content of 49% and 43%, and spheres of polystyrene randomly distributed in a polyisoprene matrix for polystyrene content of 18%[63,169]. The authors have also studied the effect of the nature of the solvent[63] using one good solvent for polystyrene (methylethyl ketone) and four good solvents of polyisoprene (cyclohexane, carbon tetrachloride, n-hexane and iso-octane). For these five solvents, electron microscopy has revealed disordered structures. These results are in contradistinction of those obtained by slow evaporation of the solvent (methylethyl ketone, dimethyl ketone and toluene) from mesophases of SI and SIS copolymers[166] of both lamellar and cylindrical type: the same type of ordered structure has always been observed in the mesophase and in the dry copolymer for the three solvents. A possible explanation of the disordered structure observed by Kawai would be a too high evaporation rate fixing the disordered structure of the dilute solution[71].

Later, toluene cast films of two SI copolymers of different molecular weight but showing electron micrographs characteristic of a lamellar structure have been studied by low-angle X-ray diffraction[173,174]. The following conclusions have been drawn: the lamellae formed by the microphase separation tend to orient themselves with their boundaries parallel to the bulk film surfaces prepared by solvent casting; the regularity of the structure decreases when the molecular weight of the copolymer increases; the annealing of the cast specimen affects the regularity and the thickness of the lamellae[173]; the interfacial thickness is rather sharp, about 3% of the total thickness of a sheet[174]. The interfacial thickness has also been studied for copolymers with spherical microdomains of polyisoprene[175] by the X-ray diffraction method previously applied to lamellar domains[174].

Kawai *et al.*[64] have also tried to relate the composition of SIS copolymers to their morphology and mechanical properties. Polystyrene spheres were found dispersed in a polyisoprene matrix for a polystyrene content of 9.5%, slightly curved polystyrene rods arranged nearly parallel in the polyisoprene matrix for a polystyrene content of 23%, a rather disordered lamellar structure for a polystyrene content of 47%, polyisoprene domains of various shapes and orientations in a polystyrene matrix for a polystyrene content of 72%. Kawai *et al.* have also observed a systematic change in the stress-strain behaviour with the copolymer composition: a change ranging from the behaviour of a soft rubber vulcanizate to that of a carbon-filled rubber vulcanizate and finally to that of a hard but toughened plastics exhibiting a well defined yield phenomenon when the polystyrene content in the copolymer increases.

To explain the existence of three types of domain structures (spherical, rodlike, and lamellar) in SI, SIS, and ISI block copolymers cast from dilute solution, Kawai *et al.*[63, 64, 172] have assumed the formation of micellar structures at a critical concentration during solvent casting. They have proposed an analysis of the formation of three types of domain structure and the size of the domains taking into account thermodynamic and molecular parameters such as incompatibility between the S and I blocks, total chain length and weight fraction composition of the copolymer, solvation of the blocks and temperature. They conclude that the block segments are preferentially oriented along the direction perpendicular to the interface between the two phases and they postulate that the micelles formed at a rather low concentration maintain their structure in the solid state without reorganization. During evaporation, the micelles shrink in the direction perpendicular to the interface between the domains. Spherical micelles shrink isotropically while rodlike and lamellar micelles shrink anisotropically. In the dry state, the diameter of the spherical domains is proportional to the power 2/3 of the degree of polymerization of the blocks located inside the domains and the diameter of the cylindrical domains and the thickness of the lamellae are roughly proportional to the power 1/2 of the degree of polymerization of the corresponding blocks.

The emulsifying effect of block and graft copolymers[177–186] and its important role in the formation of certain rubber-modified plastics such as high-impact polystyrene and ABS resins[187–192] is outside the scope of this review. Nevertheless, Kawai has studied the problem of comptabilization of homopolymers by block copolymers and we shall briefly sum up his results. Studying blends of styrene-isoprene block copolymers by light microscopy and electron microscopy, Kawai has shown that these copolymers are able to solubilize homopolymers if the molecular weight of the homopolymers is smaller than that of the corresponding blocks[170, 171, 174] confirming Riess's results[177–181].

Pedemonte *et al.*[38] have studied by low-angle X-ray diffraction and electron microscopy (on ultra-thin sections of the bulk material) a SIS copolymer containing

10% polystyrene (Kraton 1107) and the effect of extrusion and annealing on the morphology. They have shown that after the thermal treatment, the structure of the copolymer consists of a body-centered cubic arrangement of spherical polystyrene domains (Fig. 6). At our knowledge K 1107 is the single pure copolymer where the existence of an ordered cubic structure has been clearly demonstrated.

Mathis *et al.*[193] have studied by low-angle X-ray diffraction the extension of a SI copolymer containing 30 wt.-% polystyrene owning polystyrene cylinders arranged in a bidemensional hexagonal array and oriented by a shearing process at high temperature. They have found that the sample deforms elastically when the cylinders are oriented perpendicular to the stress, but that the structure is rapidly destroyed, when the stress is parallel to the cylinders.

In conclusion, we would like to remind that linear SI and SIS copolymers behave very similarly to SBS copolymers, see for instance[97, 99, 147, 194, 195] and references herein.

3) Star Copolymers of Styrene and Isoprene

In an investigation of the effect of chain geometry on the two phase morphology of polystyrene-polyisoprene block copolymers, Price *et al.* have studied films of AB linear, AB stars with 2, 3 or 4 branches[196, 197], graft copolymers[198–200], and ABC block terpolymers[201].

Linear SI and star $(SI)_nY$, where n = 2, 3, 4 and Y is, respectively, a bi-, tri tetrafunctional silicon compound, were synthetized by anionic polymerization in benzene solution using n-butyl-lithium as initiator and fractionated using the method of successive liquid-liquid phase separation. Ultrathin films (~50 nm thick) were prepared by evaporating solutions of the polymer in benzene on a clean mercury surface, stained with OsO_4 and examined with an electron microscope. For copolymers containing 25% polystyrene, a hexagonal structure with polystyrene cylinders in polyisoprene matrix has been observed and the values of the structural parameters have been found independent of the geometry of the molecule[196, 197]; low-angle X-ray diffraction studies performed on compression moulded sheets have confirmed the electron microscopy results[196].

Graft copolymers with a polystyrene skeleton were prepared by addition of poly(isoprenyllithium) to chloromethylated polystyrene in the presence of tetramethylethylenediamine[198] and fractionated by the method previously applied to star copolymers. Electron micrographs of films obtained by casting dilute hexane solution of grafted copolymers containing 26% polystyrene were characterized by regular hexagonal arrays of light spots on a dark blackground and were interpreted as corresponding to polystyrene spheres in a polyisoprene matrix[199].

ABC copolymers polystyrene-polyisoprene-poly(vinyl-2-pyridine)(S.I.V2P) with number-molecular average weight of 23000, 102000, and 23000 were prepared by stepwise anionic polymerization. Films obtained by solvent casting from methylcyclohexane and benzene were observed by electron microscopy after staining the polyisoprene block with osmium tetroxide or the poly(vinyl-2-pyridine) block with silver nitrate. A phase separation was observed and interpreted as resulting from domains formed by a blend of polystyrene and poly(vinyl-2-pyridine) dispersed in a polyisoprene matrix[201]. A microphase separation has also been observed for ABC copolymers of styrene, isoprene, and ethylene sulphide[202] and of α-methyl styrene, butadiene and styrene[203].

A nine armed star block copolymer of styrene and isoprene with a polystyrene content of 27% has been prepared by Fetters[204] *via* a two-stage anionic polymerization using m-divinylbenzene as coupling agent and studied by electron microscopy. Spheres of polystyrene arranged in a body centered cubic lattice have been claimed for this copolymer[204].

4) Copolymers of Isoprene and Vinyl-2- or Vinyl-4-pyridine (I.VP)

Rossi has synthetized block copolymers polyisoprene-poly(vinyl-2-pyridine) and polyisoprene-poly(vinyl-4-pyridine) of various composition and molecular weight by anionic polymerization under high vacuum[205, 208]. The polymerization in THF dilute solutions with Cumylpotassium as initiator yielded a 1,2 + 3,4-microstructure of the polyisoprene block. The polymerization in toluene solutions with sec-butyllithium as initiator yielded a 1,4-*cis*-microstructure of the polyisoprene block.

Mesophases prepared by dissolution of the copolymer in a preferential solvent for the poly(vinylpyridine) block (acrylic acid, nitromethane, dioxane, octanol, methylethyl ketone, ethyl acetate, vinyl acetate, styrene and methyl methacrylate) and dry copolymers obtained by slow evaporation of the solvent from the mesophases have been studied by low-angle X-ray diffraction[68, 163, 166, 206–208] and electron microscopy[206–208]. Copolymers of isoprene and vinylpyridine exhibit cylindrical hexagonal or lamellar structures depending upon their compositon. The influence of the nature, concentration, and polymerization of the solvent, molecular weight and composition of the copolymer, microstructure of the polyisoprene block, and position of the nitrogen atom in the vinylpyridine block on the values of the geometrical parameters of the periodic structures have been established[85, 208].

5) Block Copolymers of Isoprene and Methyl Methacrylate (I. MMA)

AB copolymers of isoprene and methyl methacrylate have been synthetized by Rossi[205]. A hexagonal structure formed by cylinders of polyisoprene in a matrix of poly(methyl methacrylate) has been demonstrated by low-angle X-ray diffraction and electron microscopy for copolymers containing about 30% of polyisoprene[209].

ABA copolymers poly(methyl methacrylate)-polyisoprene-poly(methyl methacrylate) having polyisoprene with a high vinyl content as central block have been synthetized by Gole *et al.*[210]. Dynamic mechanical properties of films of these ABA copolymers have been studied as a function of the copolymer composition, the temperature and the nature of the solvent (carbon tetrachloride, toluene, ethyl acetate, methylethyl ketone, dioxane) used for film preparation[210].

C) Block Copolymers Without Polydiene Blocks

1) Copolymers of Styrene and Vinyl-2- or 4-Pyridine

Grosius *et al.*[211–213] have synthetized by anionic polymerization in THF solution with cumylpotassium as initiator block copolymers polystyrene-poly(vinyl-2-pyridine) (S. V2P) and polystyrene-poly(vinyl-4-pyridine) (S. V4P). They have studied by low-angle X-ray diffraction the structure of the mesophases formed by fractions of these copolymers in solution in a preferential solvent for one block[211–213]. They have chosen toluene as a preferential solvent of the polystyrene block and octanol as a selective solvent of the poly(vinyl-2- or 4-pyridine) block; experimental difficulties have prevented them from resolving the structure of dry copolymers and mesomorphic gels containing less than about 15% solvent. Tables 5 and 6 sum up their results, but need some coments. Grosius *et al.* have used the hypothesis that for any copolymer composition the insoluble block vinyl-2- or 4-pyridine in the case of toluene, styrene in the case of octanol is always located inside the cylinders or the spheres. This hypothesis is rather surprising as it is well known that the matrix is

Table 5a. Structure of copolymers SV2P in solution in a preferential solvent for the PS block (toluene). Following Grosius *et al.*[211], the cylinders are filled with the insoluble PV2P blocks
L Lamellar structure.
Cyl Cylindrical structure

Copolymer	% PV2P	Copolymer concentration and structure
GP 7F2	56.5	100 > L > 70 > Cyl > 30
GP 10B3	18.8	90 > Cyl > 50
GP 10B2	27.0	85 > Cyl > 45
GP 10C3	36.7	80 > Cyl > 40
GP 10A3	58.6	100 > L > 60 > Cyl > 35

Table 5b. Structure of copolymers SV2P in solution in a selective solvent for the PV2P blocks (octanol). Following Grosius *et al.*[211], the cylinders are filled by the insoluble PS blocks
L Lamellar structure.
Cyl Cylindrical structure

Copolymer	% PV2P	Copolymer concentration and structure
GP 7F2	56.5	80 > Cyl > 25
GP 10B3	18.8	70 > Cyl > 30
GP 10B2	27.0	70 > Cyl > 30
GP 10C3	36.7	70 > Cyl > 40
GP 10A3	58.6	80 > Cyl > 40

formed by the blocks which are present in the largest amount. For instance the copolymer GP 10 B 3, which contains only 18.8% PV 2 P, and the copolymer GP 13 A 3, which contains only 16.3% PV 4 P, would have in octanol solution an inverse hexagonal structure formed by cylinders filled with the poly(vinyl pyridine) blocks swollen by the solvent. Nevertheless, the most important property of copolymers S.V2P and S.V4P is their ability to present two mesophases as a function of the solvent concentration for some copolymer compositions. When the solvent concentration increases, copolymers S.V2P may exhibit successively lamellar and cylindrical structures (Table 5a) while copolymers S.V4P may present successively cylindrical and cubic structures (Tables 6a and 6b). The reason for such behaviour has already been discussed[11,65]. Another original property of S.V2P copolymers is the ability of their cylindrical mesophases to solubilize homopolystyrene whose molecular weight is smaller than that of the polystyrene block of the copolymer[214].

Table 6a. Structure of copolymers SV4P in solution in a solvent for the PS block (toluene). Following Grosius *et al.*[212], the cylinders or the spheres are filled with the insoluble PV4P blocks
Cyl Cylindrical structure.
SP Cubic structure

Copolymer	% PV4P	Copolymer concentration and structure
GP 8F3	29.0	100 > Cyl > 70 > Sp > 45
GP 13A3	16.3	80 > Sp > 50
GP 13B3	26.8	70 > Sp > 40
GP 13A2	28.6	80 > Cyl > 60 > Sp > 35
GP 13B2	34.8	80 > Cyl > 55 > Sp > 30
GP 13C2	44.4	80 > Cyl > 35
GP 13D1	51.6	80 > Cyl > 35
GP 13E2	61.3	70 > Cyl > 30

Table 6b. Structure of copolymers SV4P in solution in a solvent for the PV4P blocks (octanol). Following Grosius *et al.*[212], the cylinders or the spheres are filled with the insoluble PS blocks
Cyl Cylindrical structure.
Sp Cubic structure

Copolymer	% PV4P	Copolymer concentration and structure
GP 12	72.4	85 > Cyl > 55 > Sp > 25
GP 13A3	16.3	80 > Cyl > 30
GP 13B3	26.8	80 > Cyl > 35
GP 13B2	34.8	80 > Cyl > 35
GP 13C2	44.4	70 > Cyl > 35
GP 13D1	51.6	75 > Cyl > 30

2) Copolymers of Vinyl-2- and Vinyl-4-pyridine

Using anionic polymerization in THF solution and with diphenylmethylsodium as initiator, Grosius *et al.*[215] have synthetized block copolymers poly(vinyl-2-pyridine)-poly(vinyl-4-pyridine). They have studied by low-angle X-ray scattering the structure of a copolymer V 2 P · V 4 P with molecular weight of the blocks 15000 and 8000 respectively. In octanol, they have found a lamellar and a cylindrical structure as a function of solvent concentration. In THF, dioxan, and benzene they have only found a cylindrical structure.

3) Copolymers of Different Alkyl Methacrylates

Ailhaud *et al.*[216] have synthetized by anionic polymerization in THF solution with diphenylmethylsodium or diphenylmethylpotassium as initiators the following block copolymers: poly(methyl methacrylate)-poly(hexyl methacrylate) (MMA-HMA), poly(methyl methacrylate)-poly(lauryl methacrylate) (MMA-LMA), poly(methyl methacrylate)-poly(octadecyl methacrylate) (MMA-OMA), poly(hexyl-methacrylate)-poly(lauryl methacrylate) (HMA-LMA).

They have studied by low-angle X-ray diffraction[217] copolymers MMA-HMA in a selective solvent (acetonitril) and in a "non-solvent" (decaline) for the PMMA blocks and copolymers MMA-LMA and MMA-OMA in a preferential solvent (ethyl acetate) and in a "non-solvent" (octanol) for the PMMA blocks. The study in acetonitril solutions of a set of copolymers MMA-HMA with a PHMA block of constant length and with compositions in PHMA between 17 and 79% has shown that copolymers MMA/HMA exhibit, as a function of their composition in insoluble block, four of the five structures already observed for SB copolymers by X-ray diffraction and electron microscopy[37,69,71,11], namely cubic, hexagonal, lamellar and inverse hexagonal (Table 7). These copolymers may also present two structures as a function of the solvent concentration for some compositions (Table 7).

Table 7. Structures observed[217] for a set of copolymers PMMA/PHMA with a constant molecular weight of the PHMA block (8 900) in solution in a selective solvent for the PMMA block (acetonitril)
L Lamellar structure.
H Hexagonal structure.
$\bar{H}$ Inversed hexagonal structure.
S Cubic structure

Copolymer	% PHMA	Copolymer concentration and structure
A 34F4A	17	71 > H > 66; 45 > S > 35
A 34F3B	37	70 > H > 40
A 34F2D	54	75 > L > 71; 65 > H > 44
A 34F2B	62	80 > L > 55
A 34F2A	71	78 > $\bar{H}$ > 74; 69 > L > 57
A 34F1C	79	82 > $\bar{H}$ > 68

4) Other Copolymers

A number of copolymers such as poly(butylene terephtalate)-poly(tetramethylene terephtalate)[218,219], polysulphone-poly(dimethylsiloxane)[220,221], polycarbonate-polysiloxane[222–225], polyethylene-polypropylene[226] and many others[227–237] have been synthetized and studied. However, the respective papers are mainly aimed at application, and we shall not analyze them in the present review.

D) Remarks

Block copolymers may be devided into two categories depending upon the effect of the solvent concentration on the structure of their mesophases[11,65]. To the first category belong copolymers that exhibit only one mesophase as a function of solvent concentration; SB copolymers with a polybutadiene-1.2 block are representative of this class of copolymers. To the second category belong copolymers that are able to exhibit two mesophases as a function of solvent concentrations; MMA-HMA copolymers are typical examples of such copolymers.

Another possibility of classification for systems copolymer/solvent is the selectivity of the solvent[35]. In a selective solvent for one block, copolymers exhibit, depending upon the respective volumes of the blocks, five types of structures: body centered cubic (C), cylindrical hexagonal (H), lamellar (L), inverse cylindrical hexagonal ($\bar{H}$), and inverse body centered cubic (C), the matrix being always formed by the component present in larger proportions. In a non-selective solvent, only three types of structures are observed: cylindrical and inverse cylindrical structures on one hand, and cubic and inverse cubic structures on the other hand being equivalent.

VI. Copolymers with Amorphous and Crystallizable Blocks

Among copolymers with amorphous and crystallizable blocks, segmented polyurethanes form a separate class of multiblock and multiphase polymers with a broad spectrum of mechanical properties[238–260]. In segmented polyurethanes, the molecular units and sequences are commonly divided into "hard segments" and "soft segments"[238–239]. The soft segments are flexible and generally consist of polyether or polyester long chains. The hard segments are formed by groups (aromatic rings, urea and urethane groups) that are rigid and (or) capable of strong intermolecular interactions[99]. Nevertheless, segmented polyurethanes do not present regular periodic structures and are out of the scope of this review.

On the contrary, copolymers polystyrene-poly(ethylene oxide) (SEO), polybutadiene-poly(ethylene oxide) (BEO), poly(ethyl methacrylate)-poly(ethylene oxide) (EMAEO) and polystyrene-poly(ϵ-coprolactone) (SCL) exhibit well organized periodic structures. Copolymers SEO[261–266], BEO[267–270] and SCL[271] have been studied in the dry state and in a preferential solvent for each type of block, while copolymers EMAEO have only been studied in the dry state[272–274].

We shall examine the range of stability of the ordered structures of copolymers containing an amorphous polystyrene, polybutadiene or poly(ethyl methacrylate) block and acrystallizable poly(ethylene oxide) (PEO) or poly(ϵ-caprolactone) (PCL) crystallizable block and the factors that determine the existence and the geometrical parameters of such periodic structures.

A) Phase Diagrams

Phase diagrams concentration/temperature of copolymers BEO and SEO in solution in a solvent for the amorphous block (diethyl phthalate, xylene, toluene) or of the crystallizable block (acetic acid, acrylic acid, nitromethane) have been established by studying systems copolymer/solvent of different concentrations as a function of temperature by differential scanning calorimetry, polarization microscopy, and X-ray diffraction[262]. Examples of such diagrams are given in Figs. 29 a and b. They always exhibit two phases. The phase found at temperatures higher than the melting temperature of the crystallizable chains T_f has a structure hexagonal, lamellar or inverse hexagonal depending on the copolymer composition like for polymers with amorphous blocks. The phase found at temperatures lower than T_f has always a lamellar structure with crystallized and folded PEO chains, but this structure is of the LC type in the case of a solvent of the amorphous block and of the more complex LCC type in the case of a solvent of the crystallizable block. The melting temperature T_f of the PEO chains increases with the PEO content and the molecular weight of the copolymer[265–266] but decreases with solvent concentration (Figs. 29, 30) and is higher in a solvent for the amorphous block than in a solvent for the crystallizable one (Fig. 30). The stability extends over a broader concentration range for the LC structure where it coincides with the whole region of ordered structures whereas for the LCC structure (Figs. 29, 30) it covers only a part of this region.

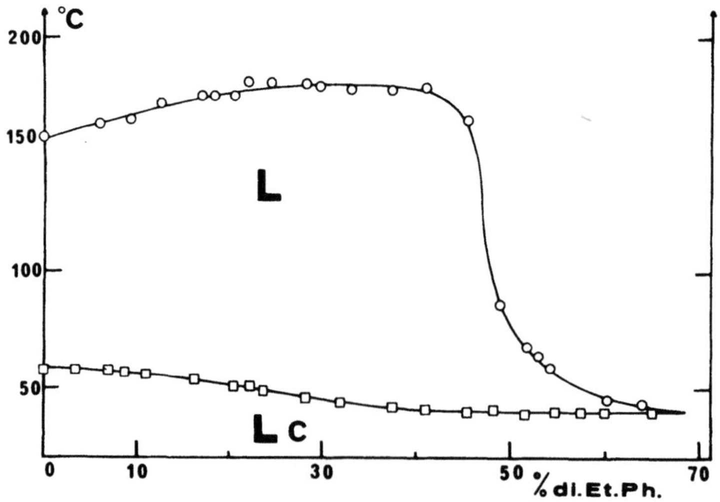

Fig. 29 (a)

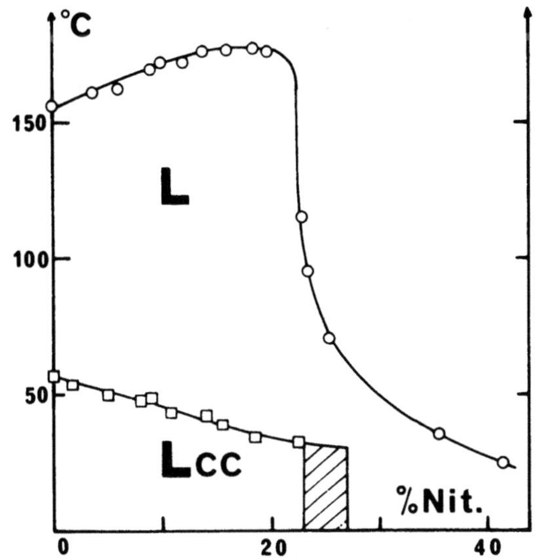

Fig. 29 (b)

Fig. 29. Phase diagrams concentration/temperature of the copolymer SEO.3 containing 59% PEO.
a: case of a selective solvent of the amorphous block[265]; b: case of a selective solvent of the crystallizable block[266];
LC: lamellar structure with crystallized PEO chains; LCC: complex lamellar structure with crystallized PEO chains; L: lamellar structure with melted PEO chains

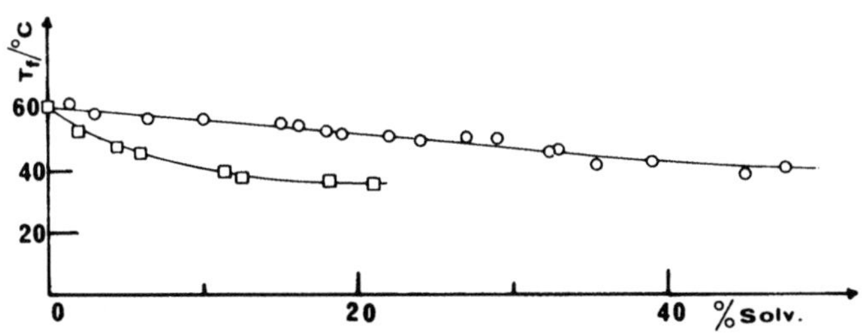

Fig. 30. Influence of the solvent concentration on the PEO melting temperature T_f for the copolymer BEO.1 in solution in xylene (○), and in acetic acid (□)[269]

B) Crystalline Structures

We shall only describe here the structure of the LC and LCC phases, because the structures found at temperatures higher than the melting temperature of the crystallizable chains are similar to that of amorphous copolymers.

In both the LC and LCC phases, the structure is lamellar and the PEO chains are crystallized as it is demonstrated by low- and wide-angle X-ray diffraction, polarization microscopy and DSC[265–266].

For the LC structure, each elementary sheet results from the superposition of two layers: a layer of thickness d_A formed by the solvated amorphous block (polystyrene or polybutadiene) and a layer of thickness d_B formed by the crystalline and folded chains (poly(ethylene oxide) or polycaprolactone)[265,269,271].

For the LCC phase, each elementary sheet also results from the superposition of two layers: a layer of thickness d_B formed by the insoluble amorphous blocks (polystyrene or polybutadiene) and a layer of thickness d_A containing the crystallized PEO chains and the solvent; but the lamella of thickness d_A has a complex structure and results from the superposition of 3 layers: the two extreme layers are constituted by the crystallized and folded PEO chains, the central layer contains the solvent[266,269].

In both the LC and LCC structures, the crystallizable chains are folded and a characteristic parameter of these two structures is the number ν of folds of the crystallizable chains; ν is given by the following formula based on simple geometrical considerations

$$\nu + 1 = \frac{2\, M_n^{cr}}{m\, d_{cr}} \tag{9}$$

In this formula, m is the weight per unit length of crystallizable PEO[275] or PCL[276] chain; d_{cr} is the thickness of the crystalline layer (d_B for the LC structure, d_{PEO} for the LCC structure); M_n^{cr} is the average number molecular weight of the crystallizable block; the factor 2 comes from the fact that the chains are crystallized in two layers[277,265,266]. Another type of folding has been proposed for EMAEO copolymers, but has not been proven[274] and is in contradiction with the results obtained for copolymers SEO and BEO in solution in a selective solvent of the PEO blocks.

C) Factors Governing Folding of Crystallizable Chains

The principal factors governing the number of folds of the crystallizable chains are: the concentration and the nature of the solvent, the crystallization temperature, the molecular weight of the two blocks, and the nature of the amorphous and crystallizable blocks. The respective influence of these factors has been analyzed in Refs. [261] to [271].

1) Influence of the Solvent Concentration

The effect of the solvent concentration on the geometrical parameters of the organized structures of copolymers with an amorphous and a crystallizable block is illustrated in Figs. 31 to 35. In a solvent for the amorphous block, one observes only one type of structures – a lamellar structure LC with crystallized and folded PEO or PCL chains (Fig. 31). In a solvent for the crystallizable block, one observes two structures dependent on the solvent concentration (Fig. 32). The first one is a LCC structure with crystallized and folded PEO chains (Fig. 32 and 33); the second one is a lamellar, hexagonal, or inverse hexagonal structure depending upon the copolymer composition, but in all cases the PEO chains are amorphous and swollen by the solvent (Fig. 32). When the solvent concentration increases, the geometrical parameters vary for the LC structure at first in a discontinuous and then in a continuous way (Figs. 31, 34), for the LCC structure always in a discontinous way (Figs. 32, 33, 35), for the L, H or $\overline{\text{H}}$ structures always in a continuous way (Fig. 32). For both the LC and LCC structures, each discontinuity increases the number of folds

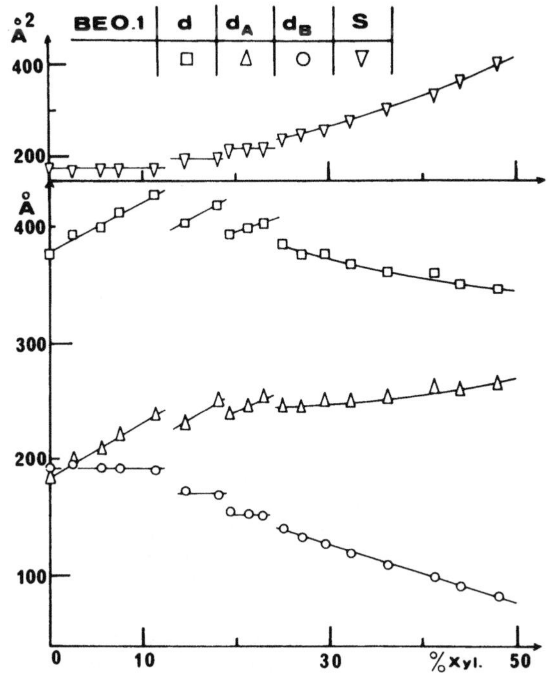

Fig. 31. Dependence of the LC structure parameters for the copolymer BEO.1 [269] on xylene concentration.
□: d = total thickness of the sheet; △: d_A = thickness of the swollen polybutadiene layer; ○: d_B = thickness of the insoluble poly(ethylene oxide) layer; ▽: Σ = specific surface

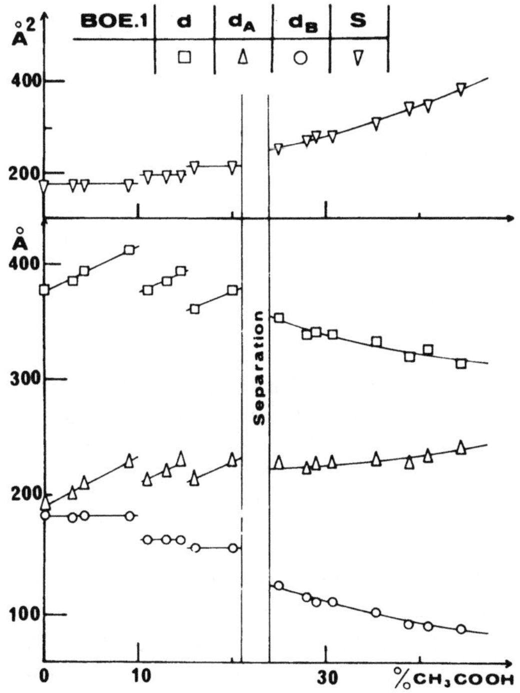

Fig. 32. Dependence of the LCC and L structure parameters for the copolymer BEO.1 [269] on acetic acid concentration.
□: d = total thickness of a sheet;
△: d_A = thickness of the layer containing the PEO chains and the solvent;
○: d_B = thickness of the insoluble polybutadiene layer; ▽: Σ = specific surface

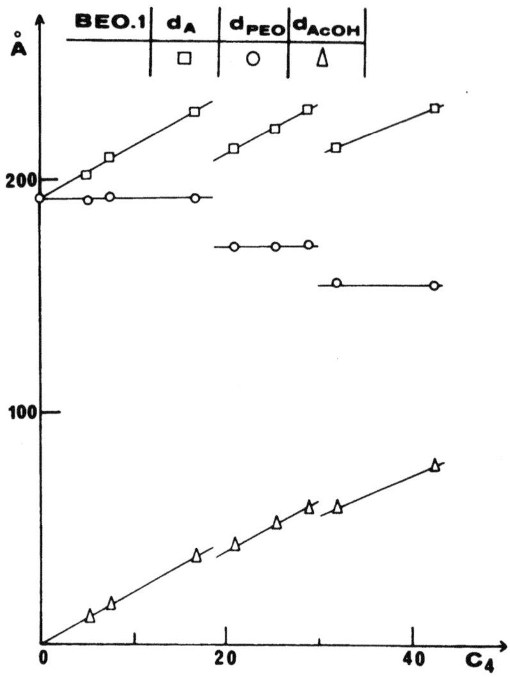

Fig. 33. Variation of the parameters $d_A = d_{(PEO+AcOH)}$, d_{PEO} and d_{AcOH} as a function of the solvent concentration C_4 for the system BEO.1/AcOH[269]

$$C_4 = \frac{\text{weight of AcOH}}{\text{weight of (PEO + AcOH)}} \%$$

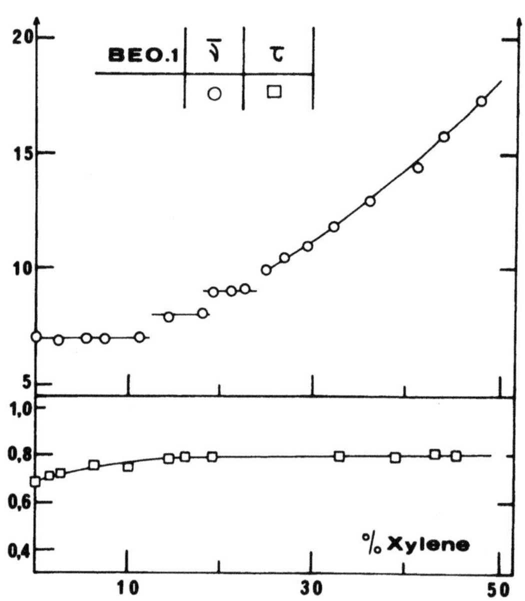

Fig. 34. Variation of the degree of crystallinity τ (□) and the average number of folds ν (○) of the PEO chains for the copolymer BEO.1 as a function of xylene concentration

of the crystallizable chains by one[265, 266, 269, 271] as is illustrated by Figs. 35 and 36. So, in both solvent types, the number of folds ν increases with solvent concentration, but for the LC structure the number of folds increases at first discontinuously and then continuously and is accompanied by a monotonous increase of the degree

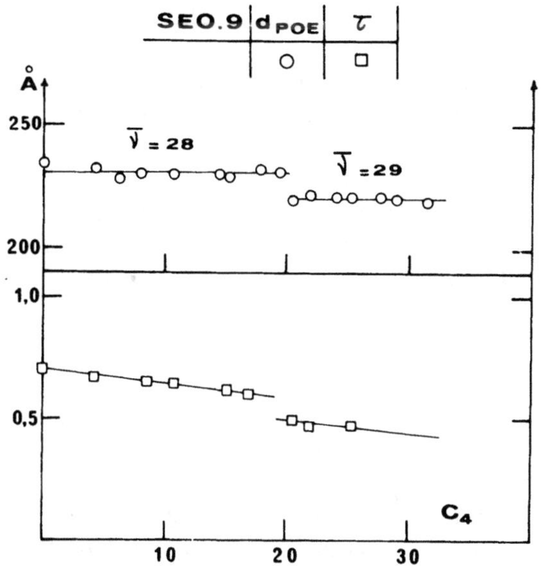

Fig. 35. Variation of the thickness d_{PEO} (○) of the crystalline PEO layer, the average degree of crystallinity τ (□) and the average number of folds ν of the PEO chains of the copolymer SEO.9 as a function of the concentration C_4

$$C_4 = \frac{\text{weight of AcOH}}{\text{weight of (PEO + AcOH)}} \%$$

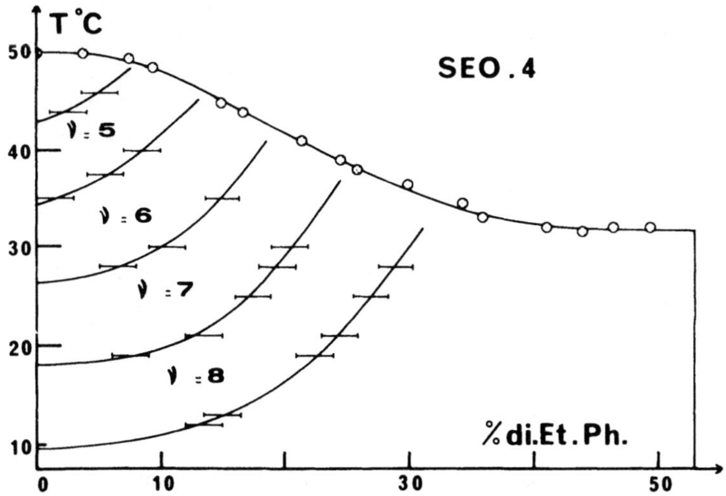

Fig. 36. Crystallization temperature/solvent concentration diagram for the system SEO.4/diethyl phthalate[265]. This diagram displays four domains in which the folding parameter ν remains constant and, respectively, equal to 8, 7, 6 and 5

of crystallinity τ (Fig. 34). For the LCC structure, the number of folds increases discontinuously and each break is accompanied by a decrease of the degree of crystallinity (Fig. 35) which provides amorphous PEO chains which can swell in the solvent located between the two crystalline PEO layers[266, 269] causing an increase in the solvent layer thickness (Fig. 33).

It has also been shown that, in the presence of a selective solvent for the amorphous blocks, for copolymers with a small content of PEO, a fractionation can occur during cristallization. But this fractionation is only observed for solvent concentrations higher than a limiting value related to the molecular weight of the PEO blocks[265].

2) Influence of the Solvent Nature

Three solvents for each type of blocks have been used: toluene, xylene, and diethyl phthalate for the amorphous blocks; acrylic acid, acetic acid, and nitromethane for the crystallizable blocks.

Copolymers SEO and BEO have the same behaviour in toluene and xylene solution; but polystyrene chains swell differently in xylene and in diethyl phthalate. The surface area occupied by a polystyrene chain at the interface between the amorphous and crystalline layer is higher in diethyl phthalate than in xylene and the PEO chains fold more frequently in diethyl phthalate than in xylene[270].

The comparison of the behaviour of systems copolymer SEO/nitromethane and copolymer SEO/acetic acid has shown that acetic acid dissolves PEO less rapidly than nitromethane and that, in the presence of the latter solvent, the PEO chains fold more easily[266].

3) Influence of the Crystallization Temperature

Systems copolymer SEO/diethyl phthalate of different concentration have been studied by DSC and low-angle X-ray diffraction as a function of the crystallization temperature. It has been found that for a fixed concentration the characteristic parameters of the LC structure vary stepwise. For instance, for a diethyl phthalate concentration of 10%, the thickness d_B of the PEO layer increases suddenly when the crystallization temperature reaches 19°, 29° and 42 °C and each jump of the d_B parameter corresponds to a decrease by one of the average number ν of folds of PEO chains[265]. An example of the phase diagram crystallization temperature/solvent concentration is given in Fig. 36 for the copolymer SEO 4 whose molecular parameters are summarized in Table 8. The degree of crystallinity of the PEO chains increases with both the crystallization temperature and solvent concentration[265].

4) Influence of the Copolymer Composition and Molecular Weight

The study by low-angle X-ray diffraction of sets of copolymers with a constant molecular weight for the crystallizable block (SEO 4 and SEO 5 or SEO 3 and SEO 8), or a constant molecular weight for the amorphous block (SEO 7, SEO 8 and SEO 9), or a constant composition but different molecular weights (SEO 3 and SEO 4), (see

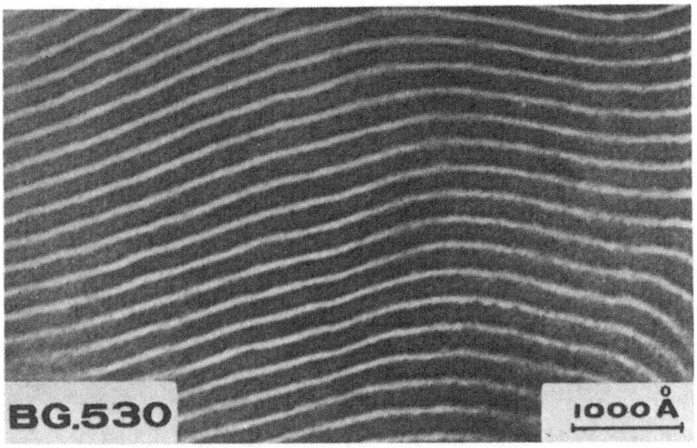

Fig. 37. Example of an electron micrograph of copolymers with a polyvinyl block and a polypeptide block. Copolymer polybutadiene-poly(benzyl-L-glutamate) BG.530 containing 33% polypeptide with polydiene chains have been stained with osmium tetroxide

Table 8. Molecular characteristics of copolymers: polystyrene-poly(ethylene oxide) (SEO) and polybutadiene-poly(ethylene oxide) (BEO)

Copolymer	$\overline{M}_n$ first block	$\overline{M}_n$ PEO	% PEO	$\bar{\nu}$
SEO 3	14 500	20 400	59	17
SEO 4	3 500	5 500	61	7
SEO 5	8 800	5 500	38.5	13
SEO 7	22 100	10 400	32	23
SEO 8	22 100	20 400	48	23
SEO 9	22 100	52 900	70.5	28
BEO 1	8 500	12 200	59	7
SEO 1	8 500	12 300	60	12

Table 8 for the molecular characteristics of the copolymers), has lead[265] to the following conclusions:

the number of folds increases with the molecular weight of the polystyrene block and the polystyrene content of the copolymer;

the number of folds is governed by the molecular weight of the polystyrene block for copolymers containing more than 50% polystyrene;

the number of folds increases with the molecular weight of the copolymer.

5) Influence of the Nature of the Blocks

The comparison of SEO and BEO copolymers with the same composition and molecular weight (SEO 1 and BEO 1 in Table 8) on one hand and the same number of folds but different molecular weight for their PEO blocks (SEO 4 and BEO 1 in Table 8) on the other hand has shown that the PEO chains are more folded when they are linked to polystyrene than when they are linked to polybutadiene. Therefore, the nature and the conformation of the amorphous block has a big influence on the folding of the crystallizable chains[269].

The comparison of copolymers SEO and SCL has shown that the nature of the crystallizable blocks governs both the number of mesophases in the copolymers and the number of folds of the crystallizable chains[278]. In a solvent for the crystallizable block as well as in a solvent for the amorphous block, SEO and BEO copolymers dependent on temperature exhibit two mesophases separated by the melting of the PEO chains. SCL copolymers contain mesophases only in a solvent for the amorphous blocks and these mesophases disappears at the melting temperature of the PCL chains, where the two types of blocks become compatible[278]. For SEO copolymers containing less than 50% PEO, the number of folds of the PEO chains is determined by the molecular weight of polystyrene which remains the dominating factor[265] at higher PEO contents. For SCL copolymers on the contrary, the number of folds of the PCL chains increases with the PCL content of the copolymer[278].

VII. Block Copolymers of Biological Interest

In biological membranes, integral proteins are amphipatic molecules: their hydrophobic moiety is embedded in the lipid bilayer and their hydrophilic moiety protrudes from the surface of the membrane[279]. So, it was interesting to prepare polymeric models of such amphipatic proteins. For that purpose, two new classes of block copolymers have been synthetized in Orleans, namely copolymers with a polyvinyl block and a polypeptide block and copolymers with a saccharide and a peptide block. We shall give some information concerning the preparation of these copolymers and then describe their structure.

A) Synthesis of Copolymers with a Polyvinyl Block and a Hydrophobic Polypeptide Block

1) Principle of the Synthesis

The method of synthesis[20–25] consists in the preparation by anionic polymerization of a monodisperse first block, followed by chemical modification of its living end in order to introduce a terminal primary amine function into the molecule; this ω aminated polymer is then used as a macromolecular initiator for the polymerization of the N-carboxy anhydride (NCA) of the desired amino acid.

2) Synthesis of Aminated Polymers

The first block (polybutadiene or polystyrene) is prepared by anionic polymerization, under high vacuum, in THF dilute solution (less than 5%), at low temperature (–70 °C) with cumylpotassium as initiator. Then, the living polymer is transformated into a hydroxylated polymer (PV–OH) by addition of ethylene oxide under vacuum, or into a carboxylated polymer (PV–COOH) by addition of carbon dioxide under vacuum.

From the hydroxylated polymer (PV–OH) one obtains at first a polymer terminated by an oxychloroformyl group (PV–O–CO–Cl) by addition of phosgene, and, at last, a polymer terminated by a primary amine function ($PV{-}NH_2$) by addition of a primary diamine[21,22].

From the carboxylated polymer (PV–COOH) one obtains an aminated polymer ($PV{-}NH_2$) by two different ways:

addition of a primary diamine in presence of a coupling adgent (dicyclohexylcarbodiimide), which is the best method of amination for polydienes[22,24];

action of sulfonyl chloride which gives a polymer terminated by a chloroformyl group (PV–COCl) followed by addition of a primary diamine, which is the best method of amination for polystyrene[24].

3) Reaction of Copolymerization

The reaction of polymerization of the NCA by the aminated polystyrene or polybutadiene is carried out in absence of moisture, at room temperature, in DMF or benzene solution, using a total concentration in NCA and aminated polyvinyl block of 2%[22,24].

4) Copolymers Prepared

The following copolymers have been prepared: polybutadiene-poly(benzyl-L-glutamate) (BG), polystyrene-poly(benzyl-L-glutamate) (SG), polystyrene-poly(cinnamyl-L-glutamate) (SC), polystyrene-poly(L-leucine) (SL), polystyrene-poly(carbobenzoxy-L-lysine) (SCK) and polybutadiene-poly(carbobenzoxy-L-lysine) (BCK).

B) Structure of Copolymers with a Polyvinyl Block and a Hydrophobic Polypeptide Block

Block copolymers with a hydrophobic polyvinyl block and a hydrophobic polypeptide block (BG, SG, SC, SL, BCK and SCK copolymers) exhibit well organized mesophases in dioxane, 1,2-dichloroethane, 2,3-dichloro 1-propene, etc., solutions. These mesophases are observed for solvent concentrations smaller than about 60% and for dry samples obtained by evaporation of the solvent at a slow rate.

1) Description of the Structure

The study of the mesophases by X-ray diffraction, electron microscopy, infrared spectroscopy and circular dichroism[20-25] has shown that the structure is always lamellar and can be described as follows: the lamellar structure consists of plane, parallel, and equidistant sheets of thickness d; each sheet results from the superposition of two layers: one of thickness d_A formed by the polyvinyl chains in a more or less random coil conformation, the other with a thickness d_B formed by the polypeptide chains in an α helix conformation, oriented perpendicular to the plane of the layers, arranged in a bidimensional hexagonal array, and generally folded.

The Fig. 37 gives an example of electron micrographs of such lamellar structure. It is characterized by alternating black (containing the polydiene blocks stained by osmium) and white (containing the polypeptide blocks) parallel stripes.

2) Factors Governing the Number of Folds

The average number of folds ν of a polypeptide chain is given by the formula

$$\nu + 1 = \overline{L}/d_B \qquad (10)$$

where $\overline{L}$ is the average length of a polypeptide chain (calculated from the number-average degree of polymerization of the polypeptide block and the projection of the distance between two polypeptide residues on the helix axis) and d_B is the thickness of the polypeptide layer[22].

The number of folds of the polypeptide chains increases with the polypeptide content of the copolymer and is higher if the polypeptide chain is linked to a polystyrene chain than to a polybutadiene chain[25,280].

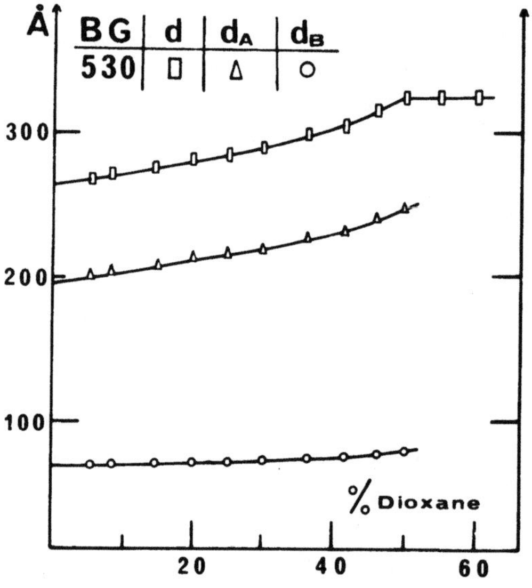

Fig. 38. Example of variation of the geometrical parameters of the lamellar structure of copolymers with a polyvinyl block and a hydrophobic polypeptide block. Copolymer BG.530 in dioxane solution[22]

3) Influence of the Solvent Concentration

When the solvent concentration increases (Fig. 38), the total thickness d of a sheet and the thickness d_A of the polyvinyl layer solvated by 70 to 80% solvent depending on the block nature[22,24] increase, while the thickness d_B of the polypeptide layer remains nearly constant; the quasi invariance of the polypeptide layer thickness is explained by the expansion of the polypeptide hexagonal lattice with swelling.

C) Structure of Copolymers with a Polyvinyl Block and a Hydrophilic Polypeptide Block

Copolymers with a hydrophobic polyvinyl block and a hydrophilic polypeptide block like polybutadiene-poly(L-lysine) (BK), polystyrene-poly(L-lysine) (SK), polybutadiene-poly(L-glutamic acid) (BE), and polystyrene-poly(L-glutamic acid) (SE) are obtained by action of HCl and HBr on copolymers BCK, SCK, BG, and SG, respectively[23].

The behaviour of copolymers SK is typical for the behaviour of copolymers with a hydrophobic polyvinyl block and a hydrophilic polypeptide block. They exhibit mesophases in water for water concentration ranging from 0 to 50% and the structure of the mesophases is lamellar[23]. The special feature of this lamellar structure consists in the conformation of the polylysine chains which are roughly to 15% in a β-chain conformation, to 35% in an α-helix conformation and to 50% in a coiled conformation[23], so that the hydrophilic block of such amphipatic copolymers has the same type of conformation as the hydrophilic part of the membrane proteins.

D) Synthesis and Structure of Copolymers with Saccharide and Peptide Blocks

In these copolymers, the saccharide block is a carbohydrate fraction (O_α or O_β) of a glycoprotein (ovomucoid) and the peptide block is a poly(benzyl-L-glutamate) block or a poly(cinnamyl-L-glutamate) block.

1) Block Copolymer Synthesis

The saccharide blocks are obtained by enzymatic (pronase) degradation of ovomucoid extracted from hen egg white[281], followed by column chromatography fractionation and purification[282]. Thus two glycoamino acids O_α and O_β are obtained. For instance, the "β fraction" (O_β) has a molecular weight 3200 and contains 16 saccharide residues: 1 residue of galactose in terminal position, 5 of mannose and 10 of N-acetyl glucoseamine[283] (Fig. 39).

The α amine function of the terminal asparagine residue of the glycoamino acids O_α or O_β has been used to initiate the polymerization of the NCA of the desired amino acids. The polymerization of the NCA has been carried out in

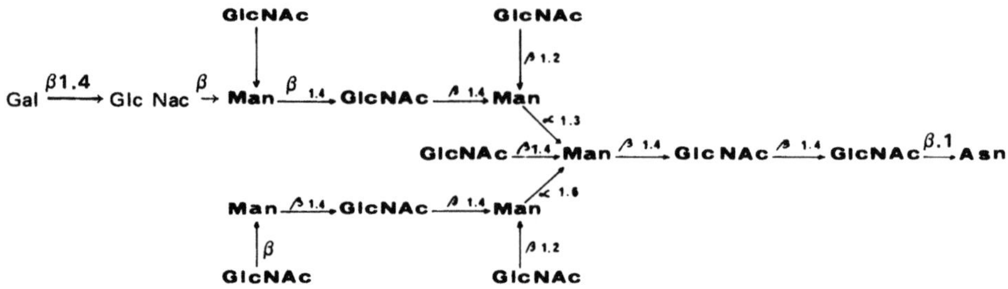

Fig. 39. Chemical structure of the glyco amino acid O_β extracted from ovomucoid

DMF in the case of the O_α glycoamino acid, and in a mixture of DMSO and DMF in the case of the O_β glycoamino acid[284].

2) Block Copolymer Structure

Copolymers with a hydrophilic saccharide block and a hydrophobic peptide block exhibit mesophases in dimethyl sulfoxide for DMSO concentrations ranging from zero to a limiting value which depends upon the nature of the carbohydrate block and the composition of the copolymer[284]; the range of stability of the mesophases increases with the saccharide content of the copolymer[285].

The study by low-angle X-ray diffraction of these mesophases provides X-ray patterns exhibiting a set of sharp lines with Bragg spacings characteristic of a layered structure. This lamellar structure results from the superposition of plane, parallel, and equidistant sheets; each sheet contains two layers: one of thickness d_A formed by the carbohydrate blocks, the other of thickness d_B formed by the polypeptide blocks; there is a partition of the solvent between the two blocks: 70% DMSO is localized in the carbohydrate layer[284]. Furthermore, the polypeptide chains are in a α-helix conformation as is demonstrated by infrared spectroscopy and X-ray diffraction and are oriented perpendicular to the plane of the sheets and assembled in a bidimensional hexagonal lattice as is revealed by X-rays[284].

The lamellar structure of saccharide-peptide block copolymers presents many analogies with the structure of copolymers with a hydrophobic polyvinyl block and a hydrophobic polypeptide block and exhibits a similar behaviour as a function of the solvent concentration. When the solvent concentration increases: the total thickness of a sheet and the thickness of the carbohydrate layer both increase while the thickness of the polypeptide layer remains nearly constant[284].

The thickness of the carbohydrate layer is nearly independant of the copolymer composition while the thickness of the polypeptide layer increases linearly with the molecular weight of the polypeptide block[285].

Acknowledgements. Thanks are expressed to Mr. P. Breton, H. Labbe and Mrs. M. Pelle for their help in the preparation of the manuscript.

VIII. References

1. Mark, H.: Textile Res. J. *23,* 294 (1953)
2. Smets, G., Hart, R.: Fortschr. Hochpolym. Forsch. *2,* 173 (1960)
3. Smets, G.: Bunsen-Diskussionstagung, Ludwigshafen (1965)
4. Kolesnikov, G. S., Yaralov, L. K.: Russian Chem. Rev. *34,* 195 (1965)
5. Gerber, G.: Makromol. Chem. *101,* 104 (1967)
6. Szwarc, M., Levy, M., Milkovich, R.: J. Amer. Chem. Soc. *78,* 2656 (1956)
7. Van Breen, A., Vlig, M.: Rubber and Plastics Age *47,* 1070 (1966)
8. Bailey, J., Bishop, E., Hendricks, W., Holden, G., Legge, N.: Rubber Age *98,* 69 (1966)
9. Holden, G., Bishop, E., Legge, N.: J. Polym. Sci., Part C, *26,* 37 (1969)
10. Bishop, E., Davison, S.: J. Polym. Sci., Part C, *26,* 59 (1969)
11. Gallot, B.: Liquid crystalline structure of block copolymers, in: Liquid crystalline order in polymers. Blumstein, A., (Ed.), New York: Academic Press, 1978, p. 191
12. Claes, P., Smets, G.: Makromol. Chem. *44–46,* 212 (1961)
13. Swarc, M.: Carbanions, living polymers and electron transfer process. New York: John Wiley, 1968
14. Szwarc, M.: Proc. Roy. Soc. (London) *A. 279,* 260 (1964)
15. Szwarc, M.: Makromol. Chem. *89,* 44 (1965)
16. Szwarc, M.: Pure Appl. Chem. *12,* 127 (1966)
17. Bywater, S., Worsfold, D. J.: Can. J. Chem. *45,* 1821 (1967)
18. Fetters, L. J.: J. Polym. Sci., Part C, *26,* 1 (1969)
19. Heuschen, J.: Thesis, University of Liège (1977)
20. Gallot, B., Perly, B., Douy, A.: IUPAC International Symposium on Macromelecules, Rio de Janeiro (1974), Preprint *D. 6–2,* p. 250
21. Perly, B., Douy, A., Gallot, B.: C. R. Acad. Sci. (Paris), *C. 279,* 1109 (1974)
22. Perly, B., Douy, A., Gallot, B.: Makromol. Chem. *177,* 2569 (1976)
23. Billot, J. P., Douy, A., Gallot, B.: Makromol. Chem. *177,* 1889 (1976)
24. Billot, J. P., Douy, A., Gallot, B.: Makromol. Chem. *178,* 1641 (1977)
25. Gallot, B., Douy, A.: IUPAC, International Symposium on Macromolecules, Dublin (1977), Vol. 1, Section II, n° 17, p. 301
26. Berger, G., Levy, M., Vofsi, D.: J. Polym. Sci., Part B, *4,* 183 (1966)
27. Skoulios, A., Finaz, G.: C. R. Acad. Sci. (Paris) *252,* 3467 (1961)
28. Finaz, G., Skoulios, A., Sadron, C.: C. R. Acad. Sci. (Paris) *253,* 265 (1961)
29. Husson, F., Mustacchi, H., Luzzati, V.: Acta Cryst. *13,* 668 (1960)
30. Luzzati, V., Mustacchi, H., Skoulios, A., Husson, F.: Acta Cryst. *13,* 660 (1960)
31. Matsuo, M.: Japan Plastics *2, 6* (1968)
32. Kawai, H., Soen, T., Inoue, T., Ono, T., Uchida, T.: Memories of the Faculty of Engineering. Kyoto University *33,* 383 (1971)
33. Lewis, P., Price, C.: Polymer *13,* 20 (1972)
34. Pedemonte, E., Dondero, G., Alfonso, G., De Candia, F.: Polymer *16,* 531 (1975)
35. Gallot, B.: Pure Appl. Chem. *38,* 1 (1974)
36. Folkes, M., Keller, A., in: Physics of glassy polymers, Haward, R. N. (ed.). London: Applied Science Publishers, 1973
37. Gallot, B., Douy, A.: Third International Liquid Crystal Conference, Berlin 1970, Preprint p. 80
38. Pedemonte, E., Turturro, A., Bianchi, U., Devetta, P.: Polymer *14,* 145 (1973)
39. Meier, D. J.: J. Polym. Sci., Part C, *26,* 81 (1969)
40. Meier, D. J.: A. C. S. Polymer Preprints *11,* 400 (1970)
41. Meier, D. J., in: Block and graft copolymers. Burke, J., Weiss, V. (Eds.). Syracuse, N.Y.: Syracuse University Press, 1973
42. Meier, D. J.: A.C.S. Polymer Preprints *15,* 171 (1974)
43. Meier, D. J.: A.C.S. Polymer Preprints *18,* 340 (1977)
44. Meier, D. J.: A.C.S. Org. Coat. Preprints *37,* 246 (1977)

45. Meier, D. J.: Polymer Colloquim. Kyoto, Japan, Sept. (1977)
46. Krause, S.: J. Polym. Sci., Part A. 2, *7*, 249 (1969)
47. Krause, S.: Macromolecules *3*, 84 (1970)
48. Bianchi, U., Pedemonte, E., Turturro, A.: J. Polym. Sci. *B. 7*, 785 (1969)
49. Bianchi, U., Pedemonte, E., Turturro, A.: Polymer *11*, 268 (1970)
50. Leary, D., Williams, M.: J. Polym. Sci. *B. 8*, 335 (1970)
51. Leary, D., Williams, M.: J. Polym. Sci., Polym. Phys. Ed. *11*, 345 (1973)
52. Leary, D., Williams, M.: J. Polym. Sci., Polym. Phys. Ed. *12*, 265 (1974)
53. Krigbaum, W., Yazgan, S., Tolbert, W.: J. Polym. Sci., Polym. Phys. Ed. *11*, 551 (1973)
54. Boehm, R., Krigbaum, W.: J. Polym. Sci., Part C, *54*, 153 (1976)
55. Helfand, E.: J. Chem. Phys. *62*, 999 (1975)
56. Helfand, E., Sapse, A.: J. Chem. Phys. *62*, 1327 (1975)
57. Helfand, E.: J. Chem. Phys. *63*, 2192 (1975)
58. Helfand, E.: Accts. Chem. Res. *8*, 295 (1975)
59. Helfand, E.: Macromolecules *8*, 552 (1975)
60. Helfand, E., Wasserman, Z.: Macromolecules *9*, 879 (1976)
61. Helfand, E., Wasserman, Z.: Polym. Eng. Sci. *17*, 582 (1977)
62. Krömer, H., Hoffman, M., Kämpf, G.: Ber. Bunsenges. Physik. Chem. *74*, 859 (1970)
63. Inoue, T., Soen, T., Hashimoto, T., Kawai, H.: J. Polym. Sci., Part A. 2, *7*, 1283 (1969)
64. Uchida, T., Soen, T., Inoue, T., Kawai, H.: J. Polym. Sci., Part. A. 2, *10*. 101 (1972)
65. Sadron, C., Gallot, B.: Makromol. Chem. *164*, 301 (1973)
66. Douy, A.: Thesis, University of Orléans (1971)
67. Douy, A., Gallot, B.: C. R. Acad. Sci. (Paris) *268*, 1218 (1969)
68. Gallot, B., Douy, A. Rossi, J.: 4th IUPAC Microsimposium, Praha, 1969, Preprint, L. 4
69. Douy, A., Gallot, B.: C. R. Acad. Sci. (Paris), *C 270*, 440 (1971)
70. Gallot, B.: 2nd Colloque international sur les méthodes analytiques par rayonnement X, Toulouse, Sept. 71, Compagnie Générale de Radiologie edit., 53 (1971)
71. Douy, A., Gallot, B.: Mol. Cryst. Liq. Cryst. *14*, 191 (1971)
72. Douy, A., Gallot, B.: in press
73. Douy, A., Gervais, M., Gallot, B.: C. R. Acad. Sci. (Paris) *C. 270*, 1646 (1970)
74. Gallot, B., Gervais, M., Douy, A.: 12th Mortonhampstead Conference of the High Polymer Research Group, 1972
75. Gervais, M., Douy, A., Gallot, B.: C. R. Acad. Sci. (Paris) *C. 276*, 391 (1973)
76. Douy, A., Gallot, B.: unpublished results
77. Hoffmann, M., Pampus, G., Marwede, G.: Kautschuk Gummi-Kunststoffe *22*, 691 (1969)
78. Kämpf, G., Hoffmann, M., Krömer, H.: Ber. Bunsenges. Physik. Chem. *74*, 851 (1970)
79. Hoffmann, M., Kämpf, G., Krömer, H., Pampus, G.: Advan. Chem. Ser. *99*, 351 (1971)
80. Kämpf, G., Krömer, H., Hoffmann, M.: Kolloid Z. Z. Polymere *247*, 820 (1971)
81. Kämpf, G., Krömer, H., Hoffmann, M.: J. Macromol. Sci. *B. 6*, 167 (1972)
82. Vanzo, E.: J. Polym. Sci., Part A. 1, *4*, 1727 (1966)
83. Bradford, E., Vanzo, E.: J. Polym. Sci., Part A. 1, *6*, 1661 (1968)
84. Douy, A., Jouan, G., Gallot, B.: C. R. Acad. Sci. (Paris) *C. 281*, 355 (1975)
85. Douy, A., Gallot, B.: Makromol. Chem., in press
86. Douy, A., Jouan, G., Gallot, B.: C. R. Acad. Sci. (Paris) *C. 282*, 221 (1976)
87. Douy, A., Jouan, G., Gallot, B.: Makromol. Chem. *177*, 2945 (1976)
88. Douy, A., Gallot, B.: Makromol. Chem. *156*, 81 (1972)
89. Douy, A., Gallot, B.: IUPAC International Symposium on Macromolecules, Leiden 1970, Preprint, Vol. 1, n° 22, p. 99
90. Douy, A., Gallot, B.: C. R. Acad. Sci. (Paris) *C. 272*, 1478 (1971)
91. Gallot, B., Douy, A.: Structures mésomorphes des copolymères sequences, in Quelques aspects de l'état solide organique. Masson (ed.) *5*, 13 (1972)
92. Douy, A., Gallot, B.: C. R. Acad. Sci. (Paris) *C 282*, 895 (1976)
93. Douy, A., Gallot, B.: C. R. Acad. Sci. (Paris) *C. 274*, 498 (1972)
94. Douy, A., Gallot, B.: IUPAC International Symposium on Macromolecules, Helsinki (1972), Preprint, Vol. 3, n° 22, p. 125
95. Douy, A., Gallot, B.: Makromol. Chem. *165*, 297 (1973)

96. Hendus, H., Illers, K., Ropte, E.: Kolloid Z. u. Z. Polymere *216–17*, 110 (1967)
97. Beecher, J., Marker, L., Bradford, R., Aggarwal, S.: J. Polym. Sci., Part C, *26*, 117 (1969)
98. Aggarwal, S., Livigni, R., Marker, L., Dudek, T.: Block and graft copolymers. New York: Syracuse Univ. Press 1970, p. 157
99. Aggarwal, S.: Polymer *17*, 938 (1976)
100. Folkes, M., Keller, A., Scalisi, F.: Kolloid Z. u. Z. Polymere *251*, 1 (1973)
101. Keller, A., Pedemonte, E., Wilmouth, F.: Nature *225*, 538 (1970)
102. Keller, A., Pedemonte, E., Wilmouth, F.: Kolloid Z. u. Z. Polymere *238*, *365* (1970)
103. Dlugosz, J., Keller, A., Pedemonte, E.: Kolloid Z. u. Z. Polymere *242*, 1125 (1970)
104. Folkes, M., Keller, A.: Polymer *12*, 222 (1971)
105. Folkes, M., Keller, A., Scalisi, F.: Polymer *12*, 793 (1971)
106. Arridge, R., Folkes, M.: J. Phys., Part D, *5*, 344 (1972)
107. Dlugosz, J., Folkes, M., Keller, A.: J. Polym. Sci., Part A. 2, *11*, 929 (1973)
108. Folkes, M., Keller, A.: J. Polym. Sci., Part A. 2, *14*, 833 (1976)
109. Folkes, M., Keller, A., Odell, J.: J. Polym. Sci., Part A. 2, *14*, 847 (1976)
110. Odell, J., Dlugosz, J., Keller, A.: J. Polym. Sci., Part A. 2, *14*, 861 (1976)
111. Arridge, R., Folkes, M.: Polymer *17*, 495 (1976)
112. Odell, J., Keller, A.: Polym. Engin. Sci. *17*, 544 (1977)
113. Pedemonte, E., Cartasegna, S., Devetta, P., Turturro, A.: Chim. Ind. Milan *55*, 861 (1973)
114. Bianchi, U., Pedemonte, E., Turturro, A., Tombini, M.: Chim. Ind. Milan *54*, 603 (1972)
115. Pedemonte, E., Turturro, A., Bianchi, U., Devetta, P.: Chim. Ind. Milan *54*, 689 (1972)
116. Turturro, A., Bianchi, U., Pedemonte, E., Devetta, P.: Chim. Ind. Milan *54*, 782 (1972)
117. Pedemonte, E., Cartasegna, S., Turturro, A.: Chim. Ind. Milan *56*, 3 (1974)
118. Pedemonte, E., Turturro, A., Dondero, G.: Br. Polymer J. *6*, 277 (1974)
119. Pedemonte, E., Alfonso, G.: Macromolecules *8*, 85 (1975)
120. Pedemonte, E., Alfonso, G., Dondero, G., de Candia, F., Araino, L.: Polymer *18*, 191 (1977)
121. Lewis, P., Price, C.: Nature *223*, 494 (1969)
122. Lewis, P., Price, C.: Polymer *12*, 258 (1971)
123. Lally, T., Price, C.: Polymer *15*, 326 (1974)
124. Fischer, E., Henderson, J.: Rubber Chem. Technol. *40*, 1373 (1967)
125. Fischer, E., Henderson, J.: J. Polym. Sci. Part C, *26*, 149 (1969)
126. Fischer, E.: J. Macromol. Sci. *A. 2*, 1285 (1968)
127. Matsuo, M., Ueno, T., Harino, H., Chigyo, S., Asai, H.: Polymer *9*, 425 (1968)
128. Matsuo, M., Sagae, S., Asai, H.: Polymer *10*, 79 (1969)
129. Kraus, G., Childers, C., Gruner, J.: J. Appl. Polym. Sci. *11*, 1581 (1967)
130. Kraus, G., Rollmann, K., Gardner, J.: J. Polym. Sci., Part A. 2, *10*, 2061 (1972)
131. Kraus, G., Rollmann, K.: J. Polym. Sci., Part A. 2, *14*, 1133 (1976)
132. Canter, N. H.: J. Polym. Sci., Part A. 2, *6*, 155 (1968)
133. Miyamoto, T., Kodama, K., Shibayama, K.: J. Polym. Sci., Part A. 2, *8*, 2095 (1970)
134. Smith, T. L., Dickie, R.: J. Polym. Sci., Part C, *26*, 163 (1969)
135. Smith, T. L., in: Block copolymers. N.Y. London: Plenum Press 1970, p. 137
136. Wilkes, G., Stein, R.: J. Polym. Sci., Part A. 2, *7*, 1525 (1969)
137. Kaeble, D.: Trans. Soc. Rheol. *15*, 235 (1971)
138. Shen, M., Kaeble, D.: J. Polym. Sci., Part B, *8*, 149 (1970)
139. Kaeble, D., Cirlin, E.: J. Polym. Sci., Part C, *43*, 131 (1973)
140. Kim, H.: Macromolecules *5*, 594 (1972)
141. Beamish, A., Hourston, D. J.: Polymer *17*, 577 (1976)
142. Beamish, A., Goldberg, R., Hourston, D. J.: Polymer *18*, 49 (1977)
143. Morton, M., McGrath, J., Juliano, P.: Polymer Sci., Part C, *26*, 99 (1969)
144. McIntyre, D., Campos Lopez, E.: Macromolecules *3*, 322 (1970)
145. Fetters, L., Meyer, B., McIntyre, D.: J. Appl. Polym. Sci. *16*, 2079 (1972)
146. Campos Lopez, E., McIntyre, D., Fetters, L.: Macromolecules *6*, 415 (1973)
147. Fedors, R.: J. Polym. Sci., Part C, *26*, 189 (1969)
148. Arnold, K., Meier, D. J.: J. Appl. Polym. Sci. *14*, 427 (1970)
149. Brown, D., Fulcher, K., Wetton, R.: J. Polym. Sci., Part B, *8*, 659 (1970)

150. Cunningham, R., Treiber, M.: J. Appl. Polym. Sci. *12*, 23 (1968)
151. Cunningham, R., Wise, M.: J. Appl. Polym. Sci. *16*, 107 (1972)
152. Cunningham, R., Auerbach, M., Floyd, W.: J. Appl. Polym. Sci. *16*, 163 (1972)
153. Conio, O., Orlandini, D., Pedemonte, E.: Rassegna Chimica, Roma *3*, 135 (1975)
154. Pedemonte, E., Dondero, G., de Candia, F., Romano, G.: Polymer *17*, 72 (1976)
155. Kraus, G., Naylor, F., Rollmann, K.: J. Polym. Sci., Part 2, *9*, 1839 (1971)
156. Kraus, G., Rollmann, K.: J. Polym. Sci., Phys. Ed. *15*, 385 (1977)
157. De Gennes, P. G.: J. Chem. Phys. *55*, 572 (1971)
158. Skoulios, A.: Macromolecules *4*, 268 (1971)
159. Gallot, B., Mayer, R., Sadron, C.: C. R. Acad. Sci. (Paris) *263*, 42 (1966)
160. Sadron, C., Gallot, B.: IUPAC International Symposium on macromolecules, Tokyo 1966
161. Gallot, B., Mayer, R., Sadron, C.: Rubber Chem. Technol. *40*, 932 (1967)
162. Gallot, B., Mayer, R., Sadron, C.: C. R. Acad. Sci. (Paris) *267*, 1298 (1968)
163. Gallot, B.: 2nd International Liquid Crystal Conference, Kent, Ohio, Aug. (1968)
164. Gallot, B., Mayer, R., Sadron, C.: 1st Republican Conference on Macromolecular Chemistry, Jassy, Romania, Sept. 1968, Preprint p. 102
165. Douy, A., Mayer, R., Rossi, J., Gallot, B.: Mol. Cryst. Liq. Cryst. *7*, 103 (1969)
166. Gallot, B., Mayer, R., Rossi, J., Douy, A.: IUPAC 4th Microsymposium, Praha, Sept. 1969, Preprint L. 3
167. Mayer, R.: Thesis, University of Strasbourg (1971)
168. Ionescu, M. L., Skoulios, A.: Makromol. Chem. *177*, 257 (1976)
169. Inoue, T., Soen, T., Kawai, H., Fukatsu, M., Kurata, M.: J. Polym. Sci., Part B, *6*, 75 (1968)
170. Inoue, T., Soen, T., Hashimoto, T., Kawai, H.: Macromolecules *3*, 87 (1970)
171. Moritani, M., Inoue, T., Motegi, M., Kawai, H.: Macromolecules *3*, 433 (1970)
172. Soen, T., Inoue, T., Miyoshi, K., Kawai, H.: J. Polym. Sci., Part A. 2, *10*, 1757 (1972)
173. Hashimoto, T., Nagatsoshi, K., Todo, A., Hasegawa, H., Kawai, H.: Macromolecules *7*, 364 (1973)
174. Hashimoto, T., Todo, A., Itoi, H., Kawai, H.: Macromolecules *10*, 377 (1977)
175. Todo, A., Uno, H., Miyoshi, K., Hashimoto, T., Kawai, H.: Polym. Engin. Sci. *17*, 587 (1977)
176. Inoue, T., Moritani, M., Hashimoto, T., Kaiwai, H.: Macromolecules *4*, 500 (1971)
177. Riess, G., Kohler, J., Tournut, C., Banderet, A.: Makromol. Chem. *101*, 58 (1967)
178. Kohler, J., Riess, G., Banderet, A.: Eur. Polym. J. *4*, 173 (1968)
179. Kohler, J., Riess, G., Banderet, A.: Eur. Polym. J. *4*, 187 (1968)
180. Periard, J., Banderet, A., Riess, G.: J. Polym. Sci., Part B, *8*, 109 (1970)
181. Periard, J., Riess, G.: Kolloid Z.u. Z. Polymere *248*, 877 (1971)
182. Periard, J., Riess, G.: Colloid and Polymer Sci. *253*, 362 (1975)
183. Marti, S., Nervo, J., Riess, G.: Progr. Coll. Polym. Sci. *58*, 114 (1975)
184. Ossenbach-Sauter, M., Riess, G.: C. R. Acad. Sci. Paris, *C. 283*, 269 (1976)
185. Reeb, R., Riess, G.: C. R. Acad. Sci. Paris, *C. 283*, 663 (1976)
186. Riess, G., Nervo, J., Rogez, D.: Polym. Engin. Sci. *17*, 634 (1977)
187. Molau, G.: J. Polym. Sci., Part A, *3*, 1267 (1965)
188. Molau, G.: J. Polym. Sci., Part A, *3*, 4235 (1965)
189. Molau, G., Kesskula, H.: J. Polym. Sci., Part A. 1, *4*, 1595 (1966)
190. Molarr, G., Wittbrodt, W.: Macromolecules *1*, 260 (1968)
191. Kato, K.: Polym. Engin. Sci. *7*, 38 (1967)
192. Bucknall, C.: Brit. Plastics *40*, 118 (1967)
193. Mathis, A., Hadziioannou, G., Skoulios, A.: Polym. Engin. Sci. *17*, 570 (1977)
194. Henderson, J., Grundy, K., Fischer, E.: J. Polym. Sci., Part C, *16*, 3121 (1968)
195. Estes, G., Cooper, S., Tobolsky, A.: Macromol. Sci., *C. 4*, 313 (1970)
196. Price, C., Lally, T., Watson, A., Woods, D., Chow, M.: Br. Polym. J. *4*, 413 (1972)
197. Price, C., Watson, A., Chow, M.: Polymer *13*, 333 (1972)
198. Price, C., Woods, D.: Polymer *14*, 82 (1973)
199. Price, C., Singleton, R., Woods, D.: Polymer *15*, 117 (1974)
200. Price, C., Woods, D.: Polymer *15*, 389 (1974)
201. Price, C., Lally, T., Stubbersfield, R.: Polymer *15*, 541 (1974)

202. Cooper, W., Hale, P., Walker, S.: Polymer *15*, 175 (1974)
203. Fielding-Russell, G., Pillai, P.: Polymer *15*, 97 (1974)
204. Le Khac Bi, Fetters, L.: Macromolecules *8*, 90 (1975)
205. Rossi, J.: Thesis, University of Orleans (1971)
206. Rossi, J., Gallot, B., Sadron, C.: C. R. Acad. Sci. (Paris), *C. 271*, 920 (1970)
207. Gallot, B., Sadron, C.: Macromolecules *4*, 514 (1971)
208. Rossi, J., Gallot, B.: Makromol. Chem., *in press* 179, 1881 (1978)
209. Rossi, J., Gallot, B.: Makromol. Chem. *177*, 2801 (1976)
210. Guyon-Gellin, J., Gole, J., Pascault, J.: J. Appl. Polym. Sci. *19*, 3173 (1975)
211. Grosius, P., Gallot, Y., Skoulios, A.: Makromol. Chem. *127*, 94 (1969)
212. Grosius, P., Gallot, Y., Skoulios, A.: Makromol. Chem. *132*, 35 (1970)
213. Grosius, P., Gallot, Y., Skoulios, A.: Europ. Polym. J. *6*, 355 (1970)
214. Skoulios, A., Helffer, P., Gallot, Y., Selb, J.: Makromol. Chem. *148*, 305 (1971)
215. Grosius, P., Gallot, Y., Skoulios, A.: Makromol. Chem. *136*, 191 (1970)
216. Ailhaud, H., Gallot, Y., Skoulios, A.: Makromol. Chem. *140*, 179 (1970)
217. Ailhaud, H., Gallot, Y., Skoulios, A.: Makromol. Chem. *151*, 1 (1972)
218. Witsiepe, W.: in Polymerization reaction and new polymers. Platzer, N.A.J. (ed.) Washington: Am. Chem. Soc., p. 39 (1973)
219. Brown, M., Witsiepe, W.: Rubber Age *104*, 35 (1972)
220. Noshay, A., Metzner, M., Merriam, C.: Polym. Prepr. *12*, 1 (1971)
221. Robeson, L., Noshay, A., Metzner, M., Merriam, C.: Angew. Makromol. Chem. *29/30*, 47 (1973)
222. Kambour, R.: J. Polym. Sci., Part B, *7*, 573 (1969)
223. Le Grand, D.: J. Polym. Sci., Part B, *7*, 579 (1969)
224. Le Grand, D.: J. Polym. Sci., Part B, *9*, 145 (1971)
225. Kambour, R.: in Block polymers. Aggarwal, S. (ed.). New York: Plenum Press 1970, p. 263
226. Kontos, E., Esterbrook, E., Gilbert, R.: J. Polym. Sci. *61*, 69 (1962)
227. Morton, M., Mikesell, S.: J. Macromol. Sci. *A. 7*, 1391 (1973)
228. Kuo, C., McIntyre, D.: J. Polym. Sci., Part A. 2, *13*, 1543 (1975)
229. Morton, M., Kammereck, R., Fetters, L.: Macromolecules *4*, 11 (1971)
230. Hale, P., Pope, G.: Europ. Polym. J. *11*, 677 (1975)
231. Falk, J., Schlott, R.: Angew. Makromol. Chem. *21*, 17 (1972)
232. Grezlak, J., Wilkes, G.: J. Appl. Polym. Sci. *19*, 769 (1975)
233. Hsuie, G., Yasukawa, G., Murakami, K.: Makromol. Chem. *139*, 285 (1970)
234. Kawai, T., Shiazaki, S., Sonoda, S., Nakagawa, H., Matsumoto, T., Maeda, H.: Makromol. Chem. *128*, 252 (1969)
235. Ono, T., Minamiguchi, H., Soen, T., Kawai, H.: Kolloid Z.u.Z. Polymere *250*, 394 (1972)
236. Soen, T., Ono, T., Yamashita, K., Kawai, H.: Kolloid Z.u.Z. Polymere *250*, 459 (1972)
237. Soen, T., Shimomura, M., Uchida, T., Kawai, H.: Colloid Polym. Sci. *252*, 933 (1974)
238. Bonart, R.: J. Macromol. Sci., *B. 2*, 115 (1968)
239. Bonart, R., Morhitzer, L., Hentze, G.: J. Macromol. Sci., *B. 3*, 337 (1969)
240. Saunders, J., Frish, K.: Polyurethanes: chemistry and technology, New York: Interscience-Wiley, 1962
241. Trappe, G., in: Advances in polyurethane technology. Buist, J., Gudgeon, H. (ed.). New York: Interscience-Wiley (1968) Chs 2 and 3
242. Wright, P., Cumming, A.: Solid polyurethane elastomers. London: MacLaren, 1969, Chapter 2
243. Puett, D.: J. Polym. Sci., *A. 25*, 839 (1967)
244. Becker, G., Oherst, H.: Kolloid Z. *152*, 1 (1957)
245. Ferguson, J., Hourston, D., Meredith, R., Patsavaudis, D.: Eur. Polym. J. *8*, 369 (1972)
246. Ferguson, J., Patsavaudis, D.: Eur. Polym. J. *8*, 385 (1972)
247. Schneider, N., Sung, C., Malton, R., Illinger, J.: Macromolecules *8*, 62 (1975)
248. Clough, S., Schneider, N., King, A.: J. Macromol. Sci., *B. 2*, 641 (1968)
249. Clough, S., Schneider, N.: J. Macromol. Sci., *B. 2*, 553 (1968)
250. Estes, G., Seymour, R., Huch, D., Cooper, S.: Polym. Engin. Sci. *9*, 383 (1969)
251. Kimura, I., Ishihara, H., Ono, H., Yoshihara, N., Nomura, S., Kawai, H.: Macromolecules *7*, 355 (1974)

252. Seymour, R., Estes, G., Cooper, S.: Macromolecules *3*, 579 (1970)
253. Kimura, I., Ishihara, N., Ono, H., Yoshihara, N., Nomura, S., Kawai, H.: Macromolecules *7*, 355 (1974)
254. Wilkes, C., Yusek, C.: J. Macromol. Sci., *B. 7*, 157 (1973)
255. Samuels, S., Wilkes, G.: J. Polymer Sci., Part C, *43*, 149 (1973)
256. Seymour, R., Allegrezza, A., Cooper, S.: Macromolecules *6*, 896 (1973)
257. Ng, H., Allegrezza, A., Seymour, R., Cooper, S.: Polymer *14*, 255 (1973)
258. Allegrezza, A., Seymour, R., Cooper, S.: Polym. Prep. *15*, 631 (1974)
259. Chang, Y., Wilkes, C.: J. Polym. Sci., Part A. 2, *13*, 455 (1975)
260. Sung Chong-Sook, P., Schneider, N.: Macromolecules *8*, 68 (1975)
261. Gervais, M., Gallot, B.: C. R. Acad. Sci. (Paris), *C. 270*, 784 (1970)
262. Gervais, M., Douy, A., Gallot, B.: Mol. Cryst. Liq. Cryst. *13*, 289 (1971)
263. Gervais, M., Jouan, G., Gallot, B.: C. R. Acad. Sci. (Paris), *C. 275*, 797 (1972)
264. Gervais, M., Jouan, G., Gallot, B.: C. R. Acad. Sci. (Paris), *C. 275*, 1243 (1972)
265. Gervais, M., Gallot, B.: Makromol. Chem. *171*, 157 (1973)
266. Gervais, M., Gallot, B.: Makromol. Chem. *174*, 193 (1973)
267. Gervais, M., Gallot, B.: IUPAC, International Symposium on macromolecules, Rio de Janeiro (1974), Preprint, Section B, n° 72, p. 128
268. Gervais, M., Jouan, G., Gallot, B.: C. R. Acad. Sci. (Paris), *C. 282*, 919 (1976)
269. Gervais, M., Gallot, B.: Makromol. Chem. *178*, 1577 (1977)
270. Gervais, M., Gallot, B.: Makromol. Chem. *178*, 2071 (1977)
271. Herman, J. J., Jerome, R., Teyssie, P., Gervais, M., Gallot, B.: Makromol. Chem. *179*, 1111 (1978)
272. Seow, P., Gallot, Y., Skoulios, A.: Makromol. Chem. *176*, 3153 (1975)
273. Seow, P., Gallot, Y., Skoulios, A.: Makromol. Chem. *177*, 177 (1976)
274. Seow, P., Gallot, Y., Skoulios, A.: Makromol. Chem. *177*, 199 (1976)
275. Tadokoro, H., Chatani, Y., Yoshihara, T., Tamara, S., Murahashi, S.: Makromol. Chem. *73*, 109 (1964)
276. Chatani, Y., Okita, Y., Tadokaro, H., Yamashita, Y.: Polymer J. *5*, 555 (1970)
277. Lotz, B., Kovacs, A., Bessett, G., Keller, A.: Kolloid Z. Polymere *209*, 115 (1966)
278. Herman, J. J., Gallot, B.: paper in preparation
279. Singer, S., Nicolson, G.: Science *175*, 720 (1972)
280. Douy, A., Gallot, B.: paper in preparation
281. Lineweaver, H., Murray, C.: J. Biol. Chem. *171*, 565 (1947)
282. Monsigny, M.: Thesis, University of Lille (1968)
283. Bayard, B.: Thesis, University of Lille (1974)
284. Douy, A., Gallot, B.: Makromol. Chem. *178*, 1595 (1977)
285. Douy, A., Gallot, B.: paper in preparation

Received April 3, 1978, June 29, 1978
K. Dušek (editor)

Author Index Volumes 1–29

Advances in Polymer Science

Fortschritte der Hochpolymeren-Forschung

Editors: H.-J. Cantow, G. Dall'Asta, K. Dušek, J.D. Ferry, H. Fujita, M. Gordon, W. Kern, G. Natta, S. Okamura, C.G. Overberger, T. Saegusa, G.V. Schulz, W.P. Slichter, J.K. Stille

Volume 25

Polymer Chemistry

1977. 55 figures. VII, 187 pages
ISBN 3-540-08389-8

Contents:

B.E. Davydov, B.A. Krentsel: Progress in the Chemistry of Polyconjugated Systems
A. Gandini: The Behaviour of Furan Derivatives in Polymerization Reactions
Z.A. Rogovin, G.A. Gabrielyan: Chemical Modifications of Fibre Forming Polymers and Copolymers of Acrylonitrile
T. Okubo, N. Ise: Synthetic Polyelectrolytes as Models of Nucleic Acids and Esterases

Volume 26

Conformation and Morphology

1978. 61 figures. IV, 185 pages
ISBN 3-540-08677-3

Contents:

W. Holzmüller: Molecular Mobility, Deformation and Relaxation Processes in Polymers
Y. Lipatov: The Iso-Free-Volume State and Glass Transitions in Amorphous Polymers: New Development of the Theory
J.E. Herz, P. Rempp, W. Borchard: Model Networks
R. Kitamaru, F. Horii: NMR Approach to the Phase Structure of Linear Polyethylene

Volume 27

Failure in Polymers
Molecular Phenomenological Aspects

1978. 97 figures. VIII, 153 pages
ISBN 3-540-08829-6

Contents:

E.H. Andrews, P.E. Reed: Molecular Fracture in Polymers
J.G. Williams: Applications of Linear Fracture Mechanics
C.B. Bucknall: Fracture and Failure of Multiphase Polymers and Polymer Composites

Volume 28

Polymerization Reactions

1978. 35 figures. IV, 151 pages
ISBN 3-540-08885-7

Contents:

Y. Yamashita: Random and Block Copolymers by Ring-Opening Polymerization
H. Sumitomo, M. Okada: Ring-Opening Polymerization of Bicyclic Acetals, Oxalactone, and Oxalactam
J.P. Kennedy, P.D. Trivedi: Cationic Olefin Polymerization Using Alkyl Halide/Alkylaluminium Initiator Systems – I. Reactivity Studies, II. Molecular Weight Studies

Springer-Verlag
Berlin Heidelberg NewYork

GPSR Compliance

The European Union's (EU) General Product Safety Regulation (GPSR) is a set of rules that requires consumer products to be safe and our obligations to ensure this.

If you have any concerns about our products, you can contact us on ProductSafety@springernature.com

In case Publisher is established outside the EU, the EU authorized representative is:

Springer Nature Customer Service Center GmbH
Europaplatz 3
69115 Heidelberg, Germany

Batch number: 10620881

Printed by Printforce, the Netherlands